AF389456

Advances in Pillared Clays and Similar Materials: Synthesis, Characterization and Applications

Advances in Pillared Clays and Similar Materials: Synthesis, Characterization and Applications

Editors

Antonio Gil Bravo
Miguel A. Vicente

MDPI • Basel • Beijing • Wuhan • Barcelona • Belgrade • Manchester • Tokyo • Cluj • Tianjin

Editors
Antonio Gil Bravo
Science Department
Public University of Navarra
Pamplona
Spain

Miguel A. Vicente
Inorganic Chemistry
Department
University of Salamanca
Salamanca
Spain

Editorial Office
MDPI
St. Alban-Anlage 66
4052 Basel, Switzerland

This is a reprint of articles from the Special Issue published online in the open access journal *Materials* (ISSN 1996-1944) (available at: www.mdpi.com/journal/materials/special_issues/pillared_clays).

For citation purposes, cite each article independently as indicated on the article page online and as indicated below:

LastName, A.A.; LastName, B.B.; LastName, C.C. Article Title. *Journal Name* **Year**, *Volume Number*, Page Range.

ISBN 978-3-0365-4828-9 (Hbk)
ISBN 978-3-0365-4827-2 (PDF)

Contents

Article

Preparation of 1D Hierarchical Material Mesosilica/Pal Composite and Its Performance in the Adsorption of Methyl Orange

Mei Wu [1,2,*], Haifeng Han [3], Lingli Ni [1], Daiyun Song [1], Shuang Li [1], Tao Hu [1], Jinlong Jiang [1,*] and Jing Chen [1]

[1]   Faculty of Chemical Engineering, Huaiyin Institute of Technology, Key Laboratory for Palygorskite Science and Applied Technology of Jiangsu Province, Huai'an 223003, China; linglini@hyit.edu.cn (L.N.); ssddyy@shu.edu.cn (D.S.); lishuang1248361263@163.com (S.L.); hutao@hyit.edu.cn (T.H.); jingchen@hyit.edu.cn (J.C.)

[2]   Huai'an Research Center of Chemical and Advanced Materials, Dalian Institute of Chemical Physics, Chinese Academy of Sciences, No. 19 Meigao Road, Wisdom Valley, Huai'an 223005, China

[3]   Jiangsu HanbonSci. & Tech Co., Ltd., No. 1-9 Ji'xian Road, Economic Development Zone, Huai'an 223003, China; hhflicp@163.com

*   Correspondence: meiwu@hyit.edu.cn (M.W.); jljiang@hyit.edu.cn (J.J.); Tel.: +86-517-8355-9056 (M.W.)

Received: 25 December 2017; Accepted: 17 January 2018; Published: 20 January 2018

Abstract: This paper highlights the synthesis of a one-dimensional (1D) hierarchical material mesosilica/palygorskite (Pal) composite and evaluates its adsorption performance for anionic dye methyl orange (MO) in comparison with Pal and Mobile crystalline material-41 (MCM-41). The Mesosilica/Pal composite is consisted of mesosilica coated Pal nanorods and prepared through a dual template approach using cetyltrimethyl ammonium bromide (CTAB) and Pal as soft and hard templates, respectively. The composition and structure of the resultant material was characterized by a scanning electron microscope (SEM), transmissionelectron microscopy (TEM), N_2 adsorption-desorption analysis, small-angle X-Ray powder diffraction (XRD), and zeta potential measurement. Adsorption experiments were carried out with different absorbents at different contact times and pH levels. Compared with Pal and MCM-41, the mesosilica/Pal composite exhibited the best efficiency for MO adsorption. Its adsorption ratio is as high as 70.4%. Its adsorption equilibrium time is as short as 30 min. Results testify that the MO retention is promoted for the micro-mesoporous hierarchical structure and positive surface charge electrostatic interactions of the mesosilica/Pal composite. The regenerability of the mesosilica/Pal composite absorbent was also assessed. 1D morphology makes it facile to separate from aqueous solutions. It can be effortlessly recovered and reused for up to nine cycles.

Keywords: adsorption; mesosilica; methyl orange; palygorskite

---

1. Introduction

The colored wastewater generated from the extensive use of dyes has produced toxicological problems and has become a major environmental concern [1]. Among the various techniques developed for effectively removing toxic dyes, adsorption is considered a simple and facile method that is adaptive to the personal and portable use for water supply [2]. Considerable efforts have been devoted to the preparation of mesoporous silica-based (mesosilica) adsorbents because of their high surface area, ordered pore structure, and tunable pore diameters [3–5]. In addition, mesosilica adsorbents have the advantage of high thermo-stability due to their silica skeleton. However, conventional mesosilica materials are difficult to separate from aqueous solutions, especially when the particle size is at the nanoscale, at which adsorbents possess considerably larger surface areas and higher adsorption capacities [6].

In recent years, one-dimensional (1D) nano-materials, such as nanofibers, nanorods, and nanotubes, have been extensively studied for environmental applications owing to their advanced properties such as large surface to volume ratio, high efficiency, and appreciable physical and chemical properties, etc. [7–10]. 1D nano-material absorbents possess a maximum surface area owing to their nanosize diameter with a high surface-to-volume ratio [11]. At the same time, their length ranges at several micrometers, at which they can be easily separated from solution by conventional methods such as filtration and centrifugation, etc. [12]. Therefore, aside from promoting dye retention with mesoporous channels, mesosilica adsorbents with 1D morphology can solve the recovery problem in liquid samples.

1D mesosilica materials can be synthesized through several methods, such as the sol-gel process [13], spray-drying method [14], laser deposition [15], and solvent evaporation techniques [16]. However, these techniques are typically hindered by their complex processes and high-cost templates. Palygorskite (Pal) is a hydrated magnesium aluminum silicate that exists in nature as a 1D fibrous mineral. The diameter of Pal is generally in the nanosize range of approximately 10 nm, and its length ranges from several hundred nanometers to several micrometers. In view of its 1D morphology and low cost, lots of 1D structures and mesoporous materials such as amorphous carbon nanotubes and mesoporous carbon nanosheets [17,18] have been prepared using Pal as a hard template. Additionally, in our previous work, Pal was used simultaneously as hard templates and a silica source to prepare amorphous carbon nanotubes and SBA-15 mesoporous materials [19]. Usually, the clay hard template should be removed with hydrofluoric acid (HF) treatment. Nevertheless, HF is a strong toxic and corrosive acid. Sacrificial clay templates cannot be recycled, and the waste leaching solution presents another environmental hazard. Actually, Pal itself is a good candidate to remove ionic dyes due to its rough surface, relatively high surface area, and moderate cation exchange capacity [20]. Even if compared with other clay minerals, Pal has superiorities in salt resistance and a rapid hydration rate. However, Pal cannot provide abundant active adsorption sites due to its aggregate crystal bundles, micropore channels, and low specific surface area.

In the present study, a 1D mesosilica/Pal composite which consisted of mesosilica coated Pal was prepared for the removal of methyl orange (MO) from solution. The Pal hard template was retained as a component, and a micro-mesoporous hierarchical structure and 1D nanorod morphology were generated. The advantages of Pal, consisting of a 1D morphology and superior adsorption performance, were extensively and innovatively taken simultaneously as the hard template and part of absorbent in this report. The micro-mesoporous hierarchical structure of 1D mesosilica/Pal composite will improve the retention time of MO and promote the MO adsorption ratio. Furthermore, its 1D nanorod morphology makes it facile to be recovered from the solution and improves its recyclability. In comparison with Pal original clay and MCM-41 mesoporous zeolite, the mesosilica/Pal composite absorbent shows the best adsorption efficiency. Results testified that it possesses superior regenerability and can be reused for up to nine cycles in the removal of MO.

2. Experimental Section

2.1. Preparation of Mesosilica/Pal Composite

The preparation of the mesosilica/Pal composite is illustrated in Scheme 1. As shown, the fabrication process includes three steps. Firstly, the Pal original clay was purified and dissociated to the scattered Pal nanorod suspension. Secondly, the surfactant cetyltrimethyl ammonium bromide (CTAB) was used to modify the Pal nanorod, resulting in the dual-template Pal-CTAB. Thirdly, under the structure-direction effect of the Pal-CTAB, the mesosilica/Pal composite was synthesized through hydrothermal crystallization.

2.1.1. Purifying and Dissociation of Pal Original Clay

To purify Pal, 5 g of the original clay was treated with 100 mL of hydrochloride acid (1 mol/L) for 48 h. The upper white Pal cake collected through high-speed centrifugation was neutralized by repeated washing, followed by freeze-drying for 12 h. In this way, loose Pal powder was obtained. Then, a given amount of Pal powder was roll extruded five times, followed by beat and ultrasonic dispersion in water. Finally, the scattered Pal nanorod suspension was obtained.

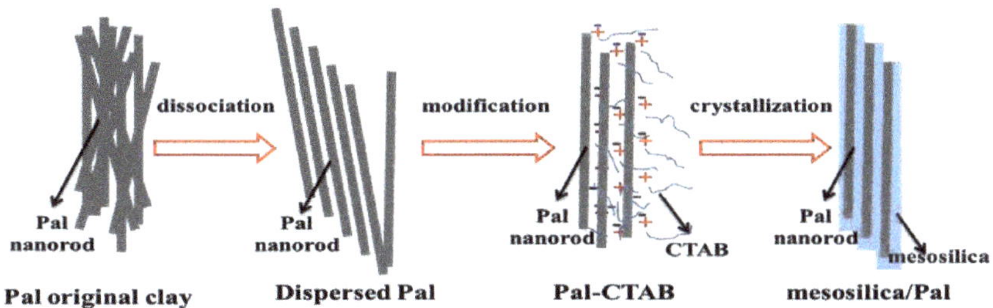

Scheme 1. Schematic illustration of mesosilica/Pal composite preparation.

2.1.2. Fabrication of Dual-Template Pal-CTAB

A total of 0.75 g of CTAB was dissolved in 10 mL deionized water and added to the scattered Pal nanorod suspension. The mixture was refluxed in an oil bath for 12 h at 80 °C. The obtained suspension was the dual-template Pal-CTAB without further treatment.

2.1.3. Preparation of Mesosilica/Pal Composite

A total of 9 g of ethanol and 4.8 g of ammonia water were added to the Pal-CTAB template. After stirring for 15 min, 1.5 g of tetraethyl orthosilicate (TEOS) was dropwise added to the solution. Then, the as-made mesosilica/Pal composite was crystallized for 48 h at 100 °C, washed, dried, and finally calcined for 5 h at 550 °C. For comparison, MCM-41 was prepared under the same condition but without the use of Pal as the hard template.

2.2. Characterization

Small-angle X-ray diffraction (XRD) was applied to characterize the long-range order of all samples. The XRD patterns were recorded on a D8-Discover diffractometer (Bruker) with Cu Kα radiation (40 kV, 40 mA). ICP-OES (Inductively Coupled Plasma Optical Emission Spectrometry) analyses were performed on a Thermo Electron IRIS Intrepid II XSP instrument (thermo fisher scientific, UK). The N_2 adsorption and desorption isotherms were measured with the Micromeritics TriStar II 3020 (Micromeritics, Norcross, GA, USA) at 77 K. The samples were out-gassed for 3 h at 300 °C under N_2 atmosphere before the measurements were recorded. The total surface area was calculated through the Brunauer–Emmett–Teller (BET) method. The total pore volume was calculated from the desorption branch of the isotherm at $P/Po = 0.99$ under the assumption of complete pore saturation. The t-plot method was adopted to evaluate the mesopore volume. The scanning electron microscopy (SEM) images of the samples were collected with an S-3000N scanning electron microscope (Hitachi, Tokyo, Japan) operated at 20 kV. The transmission electron microscopy (TEM) micrographs were obtained by using a JEM 2010 (Japan electronic materials, Tokyo, Japan) with an accelerating voltage of 200 kV. The infrared (IR) spectra of all samples were obtained with a Nicolet 5700 spectrometer (Thermo electron scientific, USA) at a resolution of 4 cm^{-1}. The zeta potential was measured with a Malvern Zetasizer Nano ZS analyzer (Malvern Instrument Ltd., UK) equipped with a multipurpose autotitrator (model MPT-2, Marvern Instruments, UK).

2.3. Evaluation of Adsorption Performance

The adsorption performance of the mesosilica/Pal composite was evaluated by conducting MO adsorption experiments in comparison with Pal and MCM-41. The contact time, pH, and recyclability of the mesosilica/Pal composite were investigated. An amount of 50 mg of the mesosilica/Pal composite was added to 20 mL MO solutions (200 mg·L^{-1}) in 100 mL conical flasks. After capping, the conical flasks were placed at room temperature and gently stirred for a specific time period at the speed of 100 rpm. They were replicated three times in the different experiments when MO adsorption was assessed. In these experiments, all assessed factors were checked in duplicate. The effect of pH on dye removal was studied over a pH range of 3–9. The initial pH of the dye solution was adjusted by adding 1 mol/L solution of HCl or NaOH. The mesosilica/Pal composite was regenerated by calcination at 500 °C for 2 h followed by separation from the waste solution.

The MO concentration (C_{MO}) and MO absorbance (A_{MO}) in the solutions were determined through a UV-vis spectrophotometer with a Hewlett Packard 8453 spectrophotometer at the wavelength of 470 nm. The adsorption ratio (q_e) was defined as the percentage removal of the adsorbate and calculated as Formula (1). The adsorption capacity (Q_{ads}, mg/g) was defined as the mass of MO adsorbed per amount of absorbents and calculated as Formula (2).

$$q_e = (C_0 - C_e)/C_0 \times 100\% \tag{1}$$

$$Q_{ads} = (C_0 - C_e) \times V_0/m_e \tag{2}$$

where C_0 and C_e are the initial and equilibrium concentrations of MO (mg/L), respectively; V_0 is the volume of solution; and m_e is the amount of absorbents.

3. Result and Discussion

3.1. Characterization of Mesosilica/Pal Composite

The micro-mesoporous hierarchical structure and 1D morphology of the prepared mesosilica/Pal composite were examined by conducting small-angle XRD measurements, SEM analysis, TEM observations, and nitrogen adsorption–desorption measurements.

As shown in Figure 1, MCM-41 exhibited well-resolved diffraction peaks at 2θ = 2.4° indexed to the (100) planes. Similarly, an obvious diffraction peak corresponding to the periodic meso-structure was observed at 2θ = 2.6° for the mesosilica/Pal composite. The appearance of these peaks implies the presence of hexagonal shaped pores in MCM-41 and the mesosilica/Pal composite [21]. However, the intensity of this relevant diffraction peak markedly decreased, as evidenced by the loss of the long-range order. This finding indicates that although periodic mesopores can be generated under the structure direction of CTAB, the Pal hard template would hamper the self-assembly of surfactant micelles and silica precursors. The slight shift of the diffraction peaks of the mesosilica/Pal composite toward the larger 2θ angles relating to MCM-41 indicated a slight shrinkage of the cell dimension [22]. The mesosilica/Pal composite only displays peaks at low angles and no any additional peaks were observed at higher angles. It demonstrates that the arrangement of atoms within the walls of the mesosilica/Pal composite is basically amorphous and the crystalline structure of Pal is collapsed totally [23].

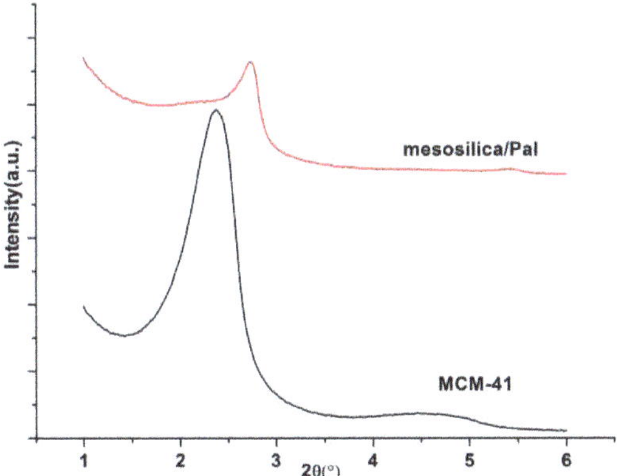

Figure 1. The small-angle XRD of the mesosilica/Pal composite and MCM-41.

The nitrogen adsorption–desorption isotherms and pore diameter distributions of the mesosilica/Pal composite and MCM-41 are shown in Figure 2. Table 1 provides the BET surface area, pore volume, and pore diameter of the mesosilica/Pal composite, MCM-41, and Pal. As shown, both the mesosilica/Pal composite and MCM-41 display a type IV isotherm, which is characteristic of ordered mesoporous materials [24]. The pore diameter of MCM-41 was distributed concentrically at 2.6 nm. The pore diameter of the mesosilica/Pal composite was presented at 2.6 and 3.9 nm, respectively, indicating a decrease in the periodic meso-structure of the mesosilica/Pal composite. This result has been confirmed by the small-angle XRD results. The mesopore at 3.9 nm in the mesosilica/Pal composite may be generated by the Pal hard template. The t-plot method was adopted to evaluate the mesopore volume. As such, the exact data of pore volume and pore diameter cannot be calculated for Pal, whose pores are nearly microporous (<2 nm). In conclusion, aside from the micropores in Pal, the mesosilica/Pal composite displayed a micro-mesoporous hierarchical structure. In addition, the BET surface of the mesosilica/Pal composite is calculated at 533.1 cm^2/g. This is lower than that of MCM-41 (937.2 cm^2/g). The probable reason for this is the small BET surface of the Pal (only 169.2 cm^2/g) used in the preparation of the mesosilica/Pal composite samples.

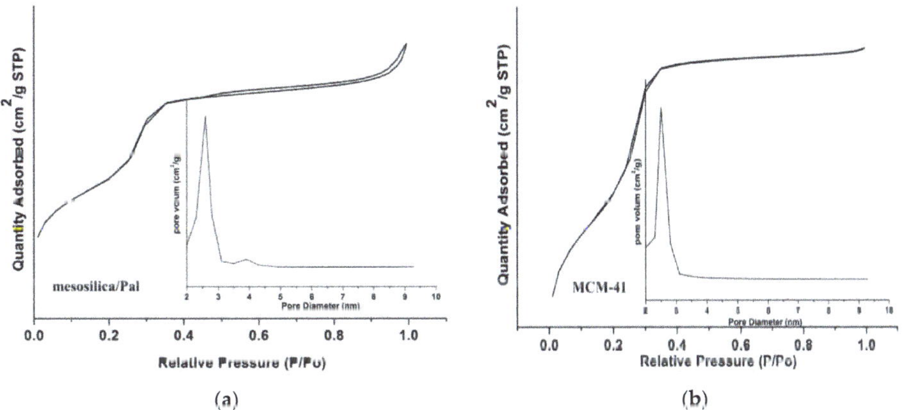

Figure 2. The N_2 adsorption and desorption isotherm curves and pore diameter distributions of the mesosilica/Pal composite and MCM-41, (**a**) mesosilica/Pal; (**b**) MCM-41.

Table 1. The physical chemistry properties of Pal, the mesosilica/Pal composite, and MCM-41.

Material	Pal	MCM-41	Mesosilica/Pal
BET surface area (cm^2/g)	169.2	937.2	533.1
Pore volume (cm^3/g)	-	0.77	0.60
Pore diameter (nm)	<2	2.6	2.6, 3.9

As presented in Figure 3, the morphologies and elements distribution of Pal, the mesosilica/Pal composite, and MCM-41 were investigated through SEM and EDX. As shown in Figure 3A, Pal crystals are formed by agglomerate nanorods in the range of 2 μm, which is a typical morphology of this mineral. The EDX analysis revealed that there are Mg, Al, K, Si, and O elements in the Pal samples, which is the general composition of a Pal mineral. In Figure 3C, MCM-41 exhibits a typical spherical particle. The particle size is uniform at approximately 200 nm. The element distribution on MCM-41 is very simple, and only includes Si and O. The morphology of the mesosilica/Pal composite is presented in Figure 3B, in which 1D short nanorods that are 200 nm in diameter and 2 μm in length are observed. Compared to Pal original clay, agglomeration for the prepared mesosilica/Pal composite is significantly improved, and all of the nanorod crystals are highly dispersed. In addition, no spherical particle can be observed in the mesosilica/Pal composite sample, indicating that mesosilica is not isolated but generated on the surface of the Pal hard template. The EDX spectra shows that mesosilica/Pal is constituted of Si, O, Mg, K, and Al elements. Except from Si and O which are from the TEOS silicate source and Pal, the remaining metal elements are all from Pal.

Figure 4A is the TEM image of the mesosilica/Pal composite under the range of 200 nm. As shown, highly dispersed Pal nanorods were wrapped with a layer of silica. To survey the morphology of the silica layer, Figure 4B, which is the amplification of Figure 4A, presents the TEM image of the mesosilica/Pal composite below 50 nm. Uniform long-range mesoporous channels were observed perpendicular to the axial direction. The average diameter of each mesoporous channel was approximately 2–3 nm, further confirming that the mesosilica covered the surface of Pal, and that a composite material was developed. Figure 4C,D presented the TEM images of MCM-41 at 100 and 50 nm, respectively. As shown, the MCM-41 prepared under the same conditions except for the application of the Pal hard template presented a long-term periodic mesoporous structure with a mesopore diameter of 2–3 nm. The obtained results were consistent with the SEM results.

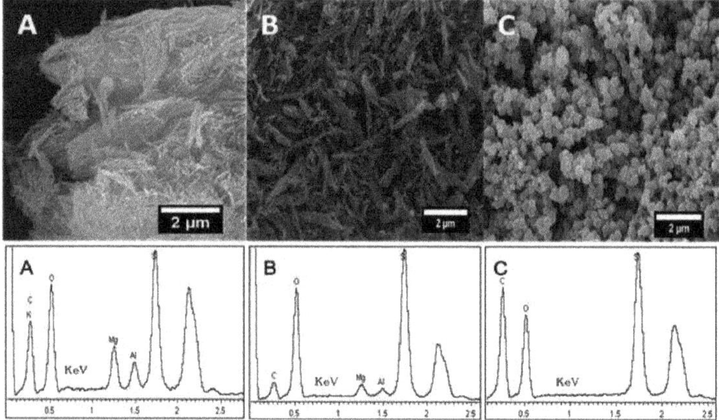

Figure 3. SEM pictures and EDX analysis of (**A**) Pal; (**B**) mesosilica/Pal composite; (**C**) MCM-41.

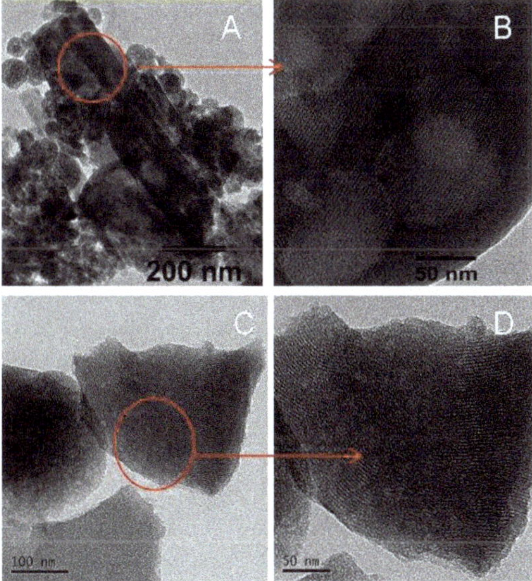

Figure 4. TEM pictures of (**A**) mesosilica/Pal composite-200 nm; (**B**) mesosilica/Pal composite-50 nm; (**C**) MCM-41-100 nm; (**D**) MCM-41-50 nm.

3.2. Methyl Orange Adsorption Property

3.2.1. MO Adsorption with Different Materials

The MO adsorption performances (including q_e, C_{MO}, and A_{MO}) of the prepared mesosilica/Pal composite, Pal, and MCM-41 samples are listed in Table 2. Simultaneously, the surface charge of each sample analyzed based on the zeta potential is also exhibited in Table 2. As shown, although MCM-41 has an excellent structure such as a high surface area and long-ranged mesopores for adsorption, its adsorption ratio was as low as 1.8%. It indicated that MCM-41 barely adsorbed MO. It is interesting that Pal, whose effective adsorption surface area was very limited, inversely had a superior adsorption ratio of 35%. Moreover, the good phenomenon is that the adsorption ratio of the mesosilica/Pal composite reached as high as 70.4%, which was much better than that of MCM-41 and Pal. The adsorption performance of these three materials can also confirmed by their digital camera images before (Scheme 2A) and after (Scheme 2B–D) adsorption, as exhibited in Scheme 2. As can be seen, the MO solution becomes nearly transparent after the adsorption of the mesosilica/Pal absorbent. It is suggested that the adsorption performance is not only determined by the pore structure and surface area, but also other factors. Thus, we attempted to analyze the causes of these results from two perspectives. On the one hand, although the mesosilica/Pal composite had less periodic mesoporous channels and a lower surface area than that of MCM-41, the micro-mesoporous hierarchical structure of the composite nanorod enhanced the retention and retarded the migration of MO. On the other hand, the adsorption performance was influenced not only by the structure, but also by the surface charge of the adsorbent. The mesoporous silica had a negative charge density because of the presence of Si–O and Si–OH groups [25]. Therefore, all of these three samples had negative zeta potentials. Among them, the mesosilica/Pal composite had the most positive zeta potential (−4.38 mV). As is known to all, MO (4-dimethylaminoazobenzene-4-sulfonic acid sodium salt) is a typical water-soluble anionic dye that tends to be attracted by positive materials. Therefore, the mesosilica/Pal composite exhibited the superior adsorption performance for MO because of its relatively higher positive surface charge.

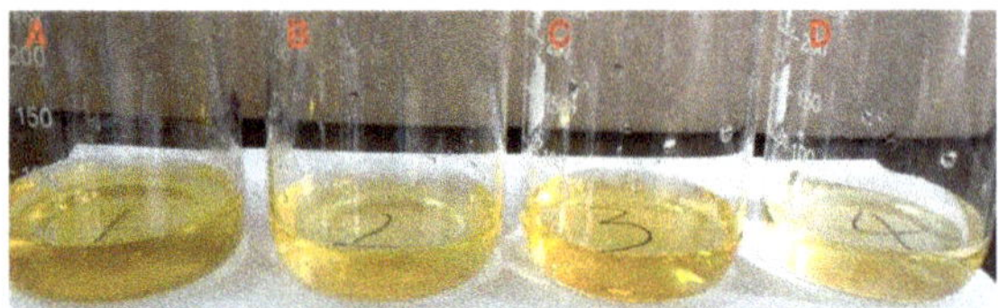

Scheme 2. Digital camera images of the materials before and after of the adsorption (**A**): The initial MO solution; (**B**): MO solution after Pal adsorption; (**C**): MO solution after MCM-41 adsorption; (**D**): MO solution after mesosilica/Pal adsorption.

Table 2. MO adsorption performance and zeta potential of the mesosilica/Pal composite, Pal, and MCM-41 (under the condition of 298 K, pH 6.0, 30 min, and an initial concentration of MO of 400 mg/L).

Materials	Zeta Potential (mV)	* A_{MO}	* C_{MO} (mg/L)	* q_e (%)	Q_{ads} (mg/g)
Pal	−20.8	0.089	178	35.0	88.80
MCM-41	−24.4	0.135	269	1.8	4.56
mesosilica/Pal	−4.38	0.040	81	70.4	178.61
Heat treated Pal	-	-	-	-	97.80 [25]
Pal-O	-	-	-	-	188.38 [26]
Mg-Al-LDHs	-	-	-	-	1471.00 [27]

* A_{MO}: MO absorbance after adsorption determined by a UV-vis spectrophotometer, the error is about ±0.002; C_{MO}: MO concentration after adsorption determined by a UV-vis spectrophotometer, the error is about ±2 mg/L; q_e: adsorption ratios of MO; Q_{ads}: adsorption capacity of MO.

We investigated the adsorption performance of some other absorbents such as 700 °C heat-treated Pal [26], acid-treated palygorskite (PAL-O) [20], and Mg–Al layered double hydroxides (LDHs) [27] proposed in the literature. The metal cations played an important bridging effect for the adsorption of anionic dyes MO. Metal cations in absorbents were used as counterions and can be adsorbed to neutralize the free $-SO_3^-$ moieties or other anionic species in the MO. For this reason, the maximum adsorption quantity of MO for Mg–Al LDHs (layered double hydroxides) (molar Mg:Al ratio of 2), much higher than general clay absorbents including the mesosilica/Pal composite reported in this work, was as high as 4.5 mmol/g (about 1471 mg/g).

To study the influency of metal cations in various absorbents, ICP-OES was applied here to investigate the metal atom content and analysis results were exhibited in Table 3. Metals such as Mg, Al, Ca, Fe, K, and Na are rich in Pal. This can be explained as the natural composition of Pal mineral. While in the MCM-41 sample, the metal atom content was lower due to the analytical pure raw material. The prepared mesosilica/Pal consisted of silica coated Pal. As a result, it has more metal atoms than MCM-41 but less than Pal. The ICP analysis results are in accordance with the above EDX results.

Table 3. The ICP-OES analysis results of various absorbents.

Materials	Al (ppm)	Ca (ppm)	Fe (ppm)	K (ppm)	Mg (ppm)	Na (ppm)	Ti (ppm)
mesosilica/Pal	15,287.64	3877.91	12,035.59	3038.33	29,620.82	6110.50	5334.48
Pal	23,241.40	8315.25	21,271.64	11,360.13	68,391.34	9435.50	3790.80
MCM-41	9726.40	23.40	−38.51	11.74	16.40	444.50	796.30

3.2.2. Effect of Contact Time on the Uptake of MO by Mesosilica/Pal Composite

To study the effect of different contact times, the adsorption ratio of the mesosilica/Pal composite was determined at varying time intervals while keeping the levels of adsorbents, pH, and temperature fixed. As presented in Figure 5, the adsorption experiments for evaluating the contact time were conducted with the time ranging from 5 min to 60 min. The adsorption ratio of MO increased rapidly in the initial 30 min and was almost unchanged for 30–40 min, indicating an equilibrium state. A 70% adsorption was reached after only 30 min, and approximately 50% of the MO was adsorbed in

10 min in the case of the mesosilica/Pal composite adsorbent. Such an adsorption performance is significantly superior to that of the other adsorbents, which typically require several hours to achieve equilibrium [28]. The rapid adsorption rate may be attributed mainly to the electrostatic interactions. This result is encouraging because the equilibrium time is an important parameter for waste water treatment. However, after the adsorption equilibrium was attained, the percentage removal of the adsorbate decreased during the next 20 min. The probable reason for this is that the initial concentration of MO significantly affected the MO adsorption. For a given adsorbent dose, the total number of available adsorption sites was fixed. The ratio of the initial mole numbers of MO to the available surface area was decreased. Thus, the percentage removal of the adsorbate decreased as the initial adsorbate concentration was reduced.

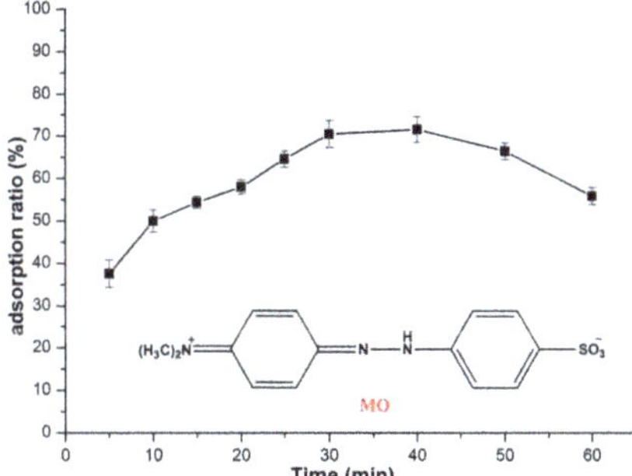

Figure 5. Effect of contact time on the uptake of MO by the mesosilica/Pal composite (under the condition of 298 K, pH 6.0, and an initial concentration of MO of 400 mg/L).

3.2.3. Effect of pH on MO Removal by Mesosilica/Pal Composite

As shown in Figure 6, the effect of pH on the MO removal by the mesosilica/Pal composite was investigated from pH = 3 to pH = 9. As can be seen, a higher adsorption ratio was found at lower pH values. The adsorption ratio is as high as 87% when the pH is 3. It is because of the protonation properties of the adsorbent with a silica skeleton [29]. Lower pH values resulted in higher hydrogen ion concentrations. The negative charges on the surface of the internal pores (as displayed in Table 3, the zeta potential of mesosilica/Pal composite −4.38 mV) were neutralized, and additional adsorption sites were developed because the surface provided a positive charge for the adsorption of the anionic MO [30].

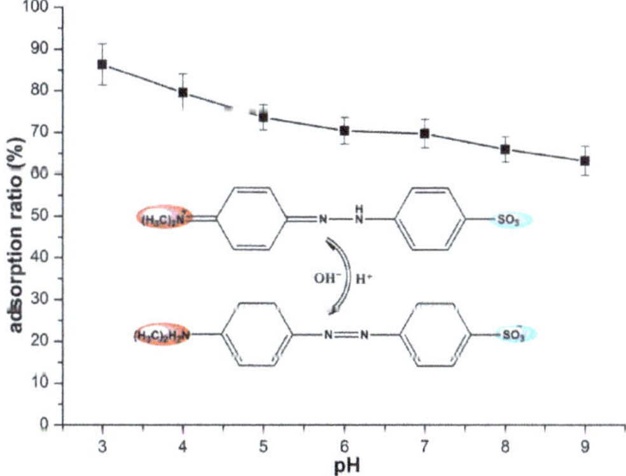

Figure 6. Effect of pH on the uptake of MO by the mesosilica/Pal composite (under the condition of 298 K, 30 min, and an initial concentration of MO of 400 mg/L).

3.2.4. The Regenerability of Mesosilica/Pal Composite

To explore the potential regenerability of the mesosilica/Pal composite as an adsorbent for MO removal, thermal treatments were conducted by calcining the exhausted adsorbents at 500 °C for 2 h in air. The specific regeneration process was illustrated in Scheme 3. The MO adsorption performances of the regenerated mesosilica/Pal composites are displayed in Figure 7. As shown, the adsorption ratio of the regenerated mesosilica/Pal composite remains almost unchange for the first five cycles. The thermal regeneration of the mesosilica/Pal composite is further feasible for nine cycles, after which the adsorption capacities of the regenerated materials suffer from progressive reductions of approximately 5% (from 70% to 65%) in comparison with that of the original mesosilica/Pal composite. The superior regeneration performance of the mesosilica/Pal composite is probably due to the high thermostability of the silica skeleton. Progressively decreasing the crystallinity of the mesosilica/Pal composite during the structural reconstruction after thermal treatment may result in decreased adsorption capacities [31].

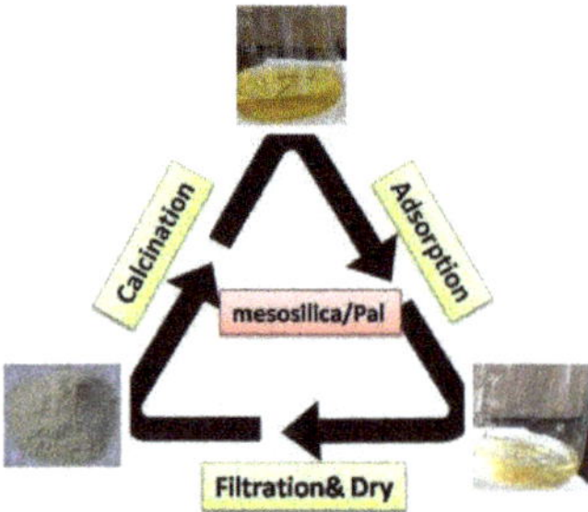

Scheme 3. Schematic illustration of the regeneration process of the mesosilica/Pal composite.

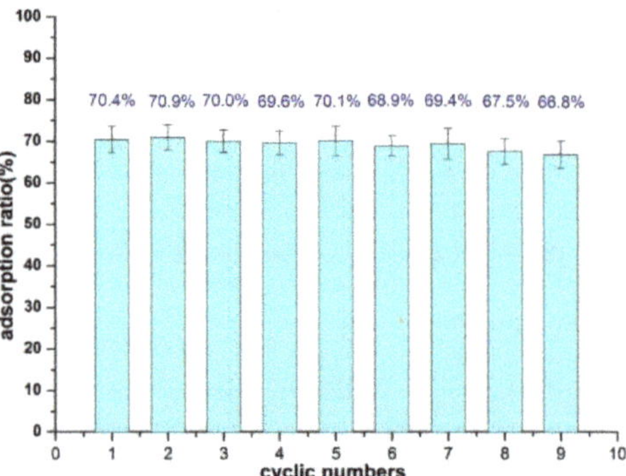

Figure 7. The regenerability of the mesosilica/Pal composite for MO adsorption (under the condition of 298 K, 30 min, pH 6.0, and an initial concentration of MO of 400 mg/L).

4. Conclusions

A novel 1D hierarchical adsorption material (mesosilica/Pal composite) was prepared through a dual template approach with CTAB and Pal as the soft and hard templates, respectively. The potential of the fabricated composite to remove MO from aqueous solution was assessed. Compared with Pal and MCM-41, the mesosilica/Pal composite was a superior adsorbent for MO removal because of its micro-mesoporous structure and relatively positive surface charge. The adsorption ratio for MO of the composite reached as high as 70.4%, and the adsorption equilibrium time was as short as 30 min.

Moreover, the 1D morphology of the mesosilica/Pal composite provided a superior regenerability, and thermal regeneration was feasible for nine cycles.

Acknowledgments: The authors sincerely acknowledge the financial support from the National Nature Science Foundation of China (Nos. 21606098, 51503072, 51574130, 11404127), Key Laboratory for Palygorskite Science and Applied Technology of Jiangsu Province (HPK201402), key scientific and technical tackle-key-problem project of Huai'an city (HAG2015011), and policy guidance program of Jiangsu province (BY2016064-01).

Author Contributions: Mei Wu, Jinlong Jiang and Jing Chen conceived and designed the experiments; Mei Wu and Jinlong Jiang interpreted the data, Haifeng Han and Lingli Ni searched literature, Daiyun Song performed the experiments; Shuang Li contributed reagents/materials/analysis tools; Haifeng Han and Tao Hu drew figures, Mei Wu wrote the paper, Lingli Ni edited the English.

Conflicts of Interest: The authors declare no conflicts of interest.

References

1. Ni, Z.-M.; Xia, S.-J.; Wang, L.-G.; Xing, F.-F.; Pan, G.-X. Treatment of methyl orange by calcined layered double hydroxides in aqueous solution: Adsorption property and kinetic studies. *J. Colloid Interface Sci.* **2007**, *316*, 284–291. [CrossRef] [PubMed]

2. Chen, A.; Li, Y.; Yu, Y.; Li, Y.; Xia, K.; Wang, Y.; Li, S. Synthesis of mesoporous carbon nanospheres for highly efficient adsorption of bulky dye molecules. *J. Mater. Sci.* **2016**, *51*, 7016–7028. [CrossRef]

3. Yokoi, T.; Kubota, Y.; Tatsumi, T. Amino-functionalized mesoporous silica as base catalyst and adsorbent. *Appl. Catal. Gen.* **2012**, *421*, 14–37. [CrossRef]

4. Lee, C.-K.; Liu, S.-S.; Juang, L.-C.; Wang, C.-C.; Lin, K.-S.; Lyu, M.-D. Application of mcm-41 for dyes removal from wastewater. *J. Hazard. Mater.* **2007**, *147*, 997–1005. [CrossRef] [PubMed]

5. Qin, Q.; Ma, J.; Liu, K. Adsorption of anionic dyes on ammonium-functionalized mcm-41. *J. Hazard. Mater.* **2009**, *162*, 133–139. [CrossRef] [PubMed]

6. Sang, Y.; Li, F.; Gu, Q.; Liang, C.; Chen, J. Heavy metal-contaminated groundwater treatment by a novel nanofiber membrane. *Desalination* **2008**, *223*, 349–360. [CrossRef]

7. Jafarzadeh, A.; Sohrabnezhad, S.; Zanjanchi, M.A.; Arvand, M. Synthesis and characterization of thiol-functionalized mcm-41 nanofibers and its application as photocatalyst. *Microporous Mesoporous Mater.* **2016**, *236*, 109–119. [CrossRef]

8. Haider, S.; Park, S.-Y. Preparation of the electrospun chitosan nanofibers and their applications to the adsorption of cu(ii) and pb(ii) ions from an aqueous solution. *J. Membr. Sci.* **2009**, *328*, 90–96. [CrossRef]

9. Li, Y.; Xiao, H.; Chen, M.; Song, Z.; Zhao, Y. Absorbents based on maleic anhydride-modified cellulose fibers/diatomite for dye removal. *J. Mater. Sci.* **2014**, *49*, 6696–6704. [CrossRef]

10. Lin, K.-S.; Cheng, H.-W.; Chen, W.-R.; Wu, C.-F. Synthesis, characterization, and adsorption kinetics of titania nanotubes for basic dye wastewater treatment. *Adsorption* **2010**, *16*, 47–56. [CrossRef]

11. Shokouhimehr, M. Magnetically separable and sustainable nanostructured catalysts for heterogeneous reduction of nitroaromatics. *Catalysts* **2015**, *5*, 534–560. [CrossRef]

12. Patil, J.V.; Mali, S.S.; Kamble, A.S.; Hong, C.K.; Kim, J.H.; Patil, P.S. Electrospinning: A versatile technique for making of 1d growth of nanostructured nanofibers and its applications: An experimental approach. *Appl. Surf. Sci.* **2017**, *423*, 641–674. [CrossRef]

13. Lu, Q.; Gao, F.; Komarneni, S.; Mallouk, T.E. Ordered sba-15 nanorod arrays inside a porous alumina membrane. *J. Am. Chem. Soc.* **2004**, *126*, 8650–8651. [CrossRef] [PubMed]

14. Bruinsma, P.J.; Kim, A.Y.; Liu, J.; Baskaran, S. Mesoporous silica synthesized by solvent evaporation: Spun fibers and spray-dried hollow spheres. *Chem. Mater.* **1997**, *9*, 2507–2512. [CrossRef]

15. Julian-Lopez, B.; Boissiere, C.; Chanenc, C.; Grosso, D.; Vadocur, S.; Miraux, S.; Duguet, E.; Sanchez, C. Mesoporous maghemite-organosilica microspheres: A promising route towards multifunctional platforms for smart diagnosis and therapy. *J. Mater. Chem.* **2007**, *17*, 1563–1569. [CrossRef]

16. Pega, S.; Boissière, C.; Grosso, D.; Azaïs, T.; Chaumonnot, A.; Sanchez, C. Direct aerosol synthesis of large-pore amorphous mesostructured aluminosilicates with superior acid-catalytic properties. *Angew. Chem. Int. Ed.* **2009**, *48*, 2784–2787. [CrossRef] [PubMed]

17. Wang, A.; Kang, F.; Huang, Z.; Guo, Z.; Chuan, X. Synthesis of mesoporous carbon nanosheets using tubular halloysite and furfuryl alcohol by a template-like method. *Microporous Mesoporous Mater.* **2008**, *108*, 318–324. [CrossRef]

18. Sun, L.; Yan, C.; Chen, Y.; Wang, H.; Wang, Q. Preparation of amorphous carbon nanotubes using attapulgite as template and furfuryl alcohol as carbon source. *J. Non-Cryst. Solids* **2012**, *358*, 2723–2726. [CrossRef]

19. Jiang, J.; Chen, Z.; Duanmu, C.; Gu, Y.; Chen, J.; Ni, L. Economical synthesis of amorphous carbon nanotubes and sba-15 mesoporous materials using palygorskite as a template and silica source. *Mater. Lett.* **2014**, *132*, 425–427. [CrossRef]

20. Yang, R.; Li, D.; Li, A.; Yang, H. Adsorption properties and mechanisms of palygorskite for removal of various ionic dyes from water. *Appl. Clay Sci.* **2018**, *151*, 20–28. [CrossRef]

21. Sohrabnezhad, S.; Jafarzadeh, A.; Pourahmad, A. Synthesis and characterization of mcm-41 ropes. *Mater. Lett.* **2018**, *212*, 16–19. [CrossRef]

22. Shang, F.; Sun, J.; Wu, S.; Yang, Y.; Kan, Q.; Guan, J. Direct synthesis of acid–base bifunctional mesoporous mcm-41 silica and its catalytic reactivity in deacetalization–knoevenagel reactions. *Microporous Mesoporous Mater.* **2010**, *134*, 44–50. [CrossRef]

23. Chuah, G.K.; Hu, X.; Zhan, P.; Jaenicke, S. Catalysts from mcm-41: Framework modification, pore size engineering, and organic–inorganic hybrid materials. *J. Mol. Catal. Chem.* **2002**, *181*, 25–31. [CrossRef]

24. Grün, M.; Unger, K.K.; Matsumoto, A.; Tsutsumi, K. Novel pathways for the preparation of mesoporous mcm-41 materials: Control of porosity and morphology. *Microporous Mesoporous Mater.* **1999**, *27*, 207–216. [CrossRef]

25. Chen, H.; Zhong, A.; Wu, J.; Zhao, J.; Yan, H. Adsorption behaviors and mechanisms of methyl orange on heat-treated palygorskite clays. *Ind. Eng. Chem. Res.* **2012**, *51*, 14026–14036. [CrossRef]

26. Darmograi, G.; Prelot, B.; Layrac, G.; Tichit, D.; Martin-Gassin, G.; Salles, F.; Zajac, J. Study of adsorption and intercalation of orange-type dyes into Mg–Al layered double hydroxide. *J. Phys. Chem. C* **2015**, *119*, 23388–23397. [CrossRef]

27. Anbia, M.; Hariri, S.A.; Ashrafizadeh, S.N. Adsorptive removal of anionic dyes by modified nanoporous silica sba-3. *Appl. Surf. Sci.* **2010**, *256*, 3228–3233. [CrossRef]

28. Chatterjee, S.; Chatterjee, S.; Chatterjee, B.P.; Das, A.R.; Guha, A.K. Adsorption of a model anionic dye, eosin y, from aqueous solution by chitosan hydrobeads. *J. Colloid Interface Sci.* **2005**, *288*, 30–35. [CrossRef] [PubMed]

29. Yao, Y.; Bing, H.; Feifei, X.; Xiaofeng, C. Equilibrium and kinetic studies of methyl orange adsorption on multiwalled carbon nanotubes. *Chem. Eng. J.* **2011**, *170*, 82–89. [CrossRef]

30. Mohammadi, N.; Khani, H.; Gupta, V.K.; Amereh, E.; Agarwal, S. Adsorption process of methyl orange dye onto mesoporous carbon material–kinetic and thermodynamic studies. *J. Colloid Interface Sci.* **2011**, *362*, 457–462. [CrossRef] [PubMed]

31. Ulibarri, M.A.; Pavlovic, I.; Barriga, C.; Hermosín, M.C.; Cornejo, J. Adsorption of anionic species on hydrotalcite-like compounds: Effect of interlayer anion and crystallinity. *Appl. Clay Sci.* **2001**, *18*, 17–27. [CrossRef]

 *materials*

Article

Preparation of Al/Fe-Pillared Clays: Effect of the Starting Mineral

Helir-Joseph Muñoz [1], **Carolina Blanco** [2], **Antonio Gil** [3], **Miguel-Ángel Vicente** [4] and **Luis-Alejandro Galeano** [1,*]

[1] Grupo de Investigación en Materiales Funcionales y Catálisis (GIMFC), Departamento de Química, Universidad de Nariño, Calle 18, Cra. 50 Campus Torobajo, 520002 Pasto, Colombia; hjmunoza@unal.edu.co

[2] Departamento de Química, Universidad Nacional de Colombia, Ciudad Universitaria, Ctra. 30 # 45-03, 111321 Bogotá, Colombia; cblancoj@unal.edu.co

[3] Departamento de Química Aplicada, Edificio de los Acebos, Universidad Pública de Navarra, Campus Arrosadia, 31006 Pamplona, Spain; andoni@unavarra.es

[4] Departamento de Química Inorgánica, Facultad de Ciencias Químicas, Universidad de Salamanca, Plaza de la Merced, s/n, 37008 Salamanca, Spain; mavicente@usal.es

* Correspondence: alejandrogaleano@udenar.edu.co; Tel.: +57-3184079325

Received: 29 October 2017; Accepted: 27 November 2017; Published: 28 November 2017

Abstract: Four natural clays were modified with mixed polyoxocations of Al/Fe for evaluating the effect of the physicochemical properties of the starting materials (chemical composition, abundance of expandable clay phases, cationic exchange capacity and textural properties) on final physicochemical and catalytic properties of Al/Fe-PILCs. The aluminosilicate denoted C2 exhibited the highest potential as starting material in the preparation of Al/Fe-PILC catalysts, mainly due to its starting cationic exchange capacity (192 meq/100 g) and the dioctahedral nature of the smectite phase. These characteristics favored the intercalation of the mixed $(Al_{13-x}/Fe_x)^{7+}$ Keggin-type polyoxocations, stabilizing a basal spacing of 17.4 Å and high increase of the BET surface (194 m^2/g), mainly represented in microporous content. According to H$_2$-TPR analyses, catalytic performance of the incorporated Fe in the Catalytic Wet Peroxide Oxidation (CWPO) reaction strongly depends on the level of location in mixed Al/Fe pillars. Altogether, such physicochemical characteristics promoted high performance in CWPO catalytic degradation of methyl orange in aqueous medium at very mild reaction temperatures (25.0 ± 1.0 °C) and pressure (76 kPa), achieving TOC removal of 52% and 70% of azo-dye decolourization in only 75 min of reaction under very low concentration of clay catalyst (0.05 g/L).

Keywords: smectite; pillared clay; keggin-like mixed Al/Fe polyoxocation; mineralogical composition; catalytic wet peroxide oxidation

1. Introduction

Clays are products of rock erosion and are found widely distributed in nature. Their chemical and textural composition varies from one place to another, depending on their geological origin and the presence of organic and inorganic impurities [1]. Bentonite clays consist mainly of phases of the smectite group, which belong to type 2:1 materials (layers of phyllosilicates formed by two tetrahedral sheets, plus one octahedral, T:O:T) condensed in a layer, and exhibit a net neutral or negative electric structure that can vary significantly between illite, smectites, vermiculites, micas, etc. [2] Smectites exhibit many interesting properties, especially for their application in adsorption and heterogeneous catalysis, originating in their peculiar physical, chemical and crystalline character. Among these properties, their cationic exchange capacity (CEC) and ability to swell in polar media make clays interesting as raw materials for modification through soft chemistry and intercalation methods. Pillaring has been one of the most widely used techniques in structural modification of this type of materials over the past 20 years [3–5].

In general, it has been documented that the success in the preparation of pillared clays strongly depends on the physical and chemical characteristics of the starting mineral. The most influential properties for this application include swellability, the net content of expandable clay minerals [4], the content of exchangeable cations as measured with the CEC, the chemical nature of such cations (in particular their radius of hydration) [6–8], the degree of isomorphic substitution in the tetrahedral and octahedral sheets, which indirectly is also related to the CEC, and the crystallinity of the smectite phases, among others [2,6,9–12]. The chemical composition [2] and distribution of transition metals active in the CWPO reaction, especially Fe and Cu, within the structural layers and extra-structural impurities are also important for the specific case of pillaring with the mixed Al/Fe system [13].

Al/Fe pillared clays have exhibited high performance activating the catalytic wet peroxidation in heterogeneous phase, and have therefore been used in the degradation of several organic compounds present in water, including emerging pollutants [6,14], phenolic compounds [8,15–17], natural organic matter [12,18] and as various toxic and bio-refractory azo-dyes [13,19,20], including methyl orange. The catalytic performance of Al/Fe-PILCs in the CWPO reaction is also closely related to the structural characteristics of the starting clay [21]. The CEC of the starting material strongly determines the ability of the Al/Fe polycations to replace the exchangeable cations originally present in the ore [8]. The amount of Fe effectively intercalated in the smectite and its specific location in the final structure of the catalyst directly affects the redox activity that the active metal may exhibit in the CWPO process, as well as its stability in the reaction medium [13]. The exchangeable cations may also play an important role as they have the ability to influence the swelling capacity of the raw material, and thus to influence the rate and the intercalation efficiency of the mixed oligomers [6]. The textural properties of the starting aluminosilicate may also be considered; the low dimensionality of the porous channels in this type of layered minerals implies that the molecules on the surface are more likely to collide with each other than they could in three dimensions. This, in turn, leads to a higher frequency of collisions and, consequently, greater reactivity [22].

Once the pillaring of the starting aluminosilicate has been carried out with the larger oligocations, the specific surface area, mainly represented in the area of micropores now available to catalyze the CWPO reaction, must substantially increase. An aspect that might negatively impact the application of a starting material in the preparation of active clay catalysts in the CWPO reaction is the presence of extra-structural Fe aggregates in the starting aluminosilicate, as these have shown to get easily leached during the catalytic action of the Al/Fe-PILCs [6]. However, in the treatment of wastewater under a continuous regime, it should not be a major obstacle. In general, the catalytic performance of these materials in the reaction-of-interest will depend on the type of iron species incorporated, its distribution, accessibility and the chemical environment of these active sites [4].

In the present work, four natural clays (denoted C1, C2, C3 and a well-known reference material BV) were, therefore, modified with a mixed intercalating $(Al/Fe)_{13}^{7+}$ solution, prepared according to a widely reported methodology [13,18,23–25]. The samples presented different: (i) clay content, (ii) distribution of clayey phases (iii) CEC, (iv) distribution of exchangeable cations (sodium, calcium or magnesium), (v) chemical composition, with presence of extra-structural iron aggregates, and (vi) textural properties. The physicochemical properties of the clays were evaluated before and after intercalation with the Al/Fe oligomers, and a correlation was made with the catalytic potential, exhibited by the final materials in the CWPO degradation of methyl orange in dilute aqueous solutions.

2. Materials and Methods

2.1. Materials

Four Colombian natural clays were employed as starting material, denoted as: class 1 (C1), class 2 (C2), class 3 (C3) and bentonite of the *Valle del Cauca* (BV). Clays (C1, C2 and C3) were selected and collected taking into account the following criteria: (i) they are all highly available, low-cost materials that have not been characterized or modified by pillaring procedures so far; (ii) ores from which aluminosilicates were extracted are currently exploited along with guaranteed long time exploitation horizon; (iii) besides, mineralogical analyses showed significant but different contents of expandable phases (smectite type), cationic exchange capacities and elemental composition, appropriate to figure out changes in Al/Fe-pillaring as a function of such physicochemical properties in the starting clays in the context of the heterogeneous Fenton, CWPO degradation of contaminants. On the other hand, the BV clay was used as a reference mineral as its use has been widely documented in academic literature in preparation of pillared clays [13,26–28]. These raw materials were refined by sedimentation of aqueous suspensions using Stokes's law, allowing the separation of the fraction with particle diameters that are lower than 2 μm. The refined materials are then symbolized by the acronyms C1-R, C2-R, C3-R and BV-R, respectively, and were modified with the mixed intercalating solution Al/Fe prepared in diluted, standard conditions. The Keggin-type mixed oligomeric precursor was prepared using $AlCl_3 \cdot 6H_2O$ (99%, Sigma-Aldrich®, St. Louis, MO, USA), $FeCl_3 \cdot 6H_2O$ (97%, Sigma-Aldrich®, St. Louis, MO, USA) and NaOH (99%, Merck®, Billerica, MA, USA), used as received. Ammonium acetate (97%, Carlo Erba®, Barcelona, Spain) was used to determine CEC.

2.2. Preparation of Pillared Clays

The source materials (C1, C2, C3 and BV) were pillared in diluted medium with the $(Al/Fe)_{13}^{7+}$ mixed system, following a standard procedure widely reported in the literature [13,18,29,30]. First, the pillaring dissolution was prepared with 0.18 mol/L of $AlCl_3 \cdot 6H_2O$ and 0.02 mol/L of $FeCl_3 \cdot 6H_2O$ solutions in an appropriate ratio to reach a final atomic ratio of 5.0% of the active metal (AMR_{Fe}) in the Al/Fe solution and total metal concentration (TMC) of 0.06 mol/L after finishing the hydrolysis stage. These concentrations have proven to be optimal for the inclusion of a major fraction of the metals being part of the Keggin Al/Fe mixed polyoxocations.

Subsequently, a 0.2 mol/L solution of NaOH was slowly added at 70 °C in sufficient quantity in order to obtain a final hydrolysis ratio (HR = $HO^-/(Al^{3+} + Fe^{3+})$) of 2.4. The resulting solution was then aged at the same temperature for 2 h, and then slowly added (drop by drop) onto a 2.0% (w/v) suspension of each clay in water under vigorous stirring. Intercalated clays were repeatedly washed with distilled water using a dialysis membrane (Sigma®, St. Louis, MO, USA), dried at 60 °C, and calcined at 500 °C for 2 h in the open air to obtain the pillared materials (C1-P, C2-P, C3 P and BV-P). A general sketch of the experimental procedure is summarized in Figure 1.

2.3. Physicochemical Characterization

Elemental analysis of the starting, extracted, and pillared aluminosilicates was performed by X-Ray Fluorescence (XRF), for which calibration curves were created using the QUANT-EXPRESS method (Fundamental Parameters) in a Bruker S8 Tiger 4 KW Wavelength Dispersive X-ray Spectrometer, with an Rh anode as an X-ray source, scintillation detector (heavy elements, from the Ti to the U), and flow (light elements, from the Na to the Sc). For this analysis, approximately 1.0 g of sample was used, sieved through a 400 mesh, and calcined at a heating rate of 3.0 °C/min up to 950 °C in order to determine the ignition losses.

The CEC of the refined and pillared clays was determined by saturation at reflux temperature using 45 mL of 2.0 mol/L ammonium acetate solution per g of solid, followed by repeated washing with distilled water and centrifugation to remove excess of ammonium ions. The content of exchanged NH_4^+ ions was then determined by the micro-Kjeldahl method and finally expressed as meq. $NH_4^+/100$ g of solid [16].

X-ray diffraction (XRD) patterns of the raw, refined (starting materials), and pillared minerals were determined using a Bruker D8 Advance diffractometer, operating at 40 kV and 30 mA with a scanning speed of 2.29 °2θ/min, employing Cu Kα radiation (λ = 0.15416 nm). The materials were analyzed by default in the range of 2.0° to 70.0°. Determinations on oriented specimens were measured in samples deposited on glass plates and dried at room temperature in a range between 2.0° and 30°.

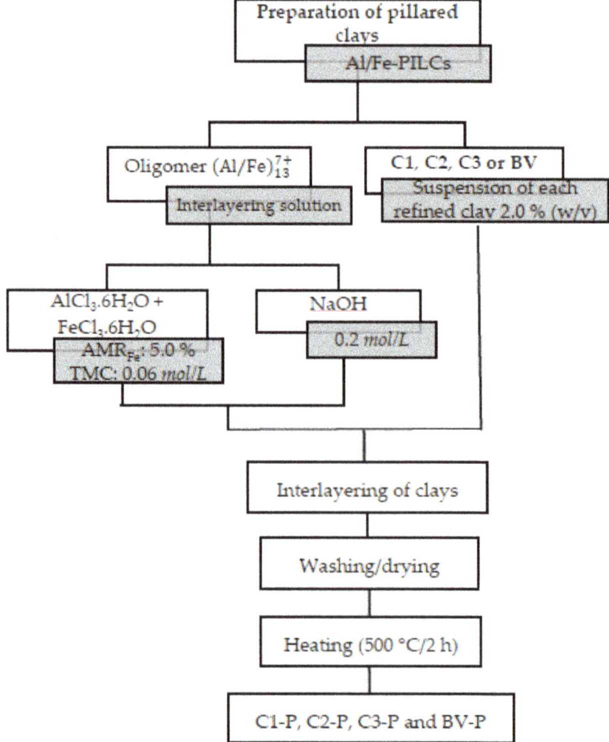

Figure 1. Experimental procedure of the preparation of pillared clays.

For the semi-quantitative determination of the smectite content in every mineral, the methodology proposed by Thorez [31] was adopted, measuring the diffractograms of oriented plates of the initial, raw aluminosilicates (C1, C2, C3 and BV), consecutively performing the expansions on each specimen with ethylene glycol and finally thermal collapse of the aluminosilicate layers (T = 400 °C/2 h). The preparation of the oriented plates was performed in the following manner: first, an extensive cationic exchange of each of the materials was carried out using a 0.5 mol/L CaCl$_2$ solution. The suspensions of each Ca^{2+}-homoionized mineral were well dispersed with ultrasound for 15 min, and then deposited with a Pasteur pipette on small glass sheets fitting the sample holder of the diffractometer. They were then allowed to dry at room temperature and the diffractograms of the oriented films measured. Afterwards, the plates were saturated with ethylene glycol, according to the methodology proposed by Moore et al. [32], in which the plates were solvated by exposing them to solvent vapor in a desiccator at room temperature for 16 h. A few minutes after the treatment's finish, the plates were removed and the diffractograms measured again under the same conditions. Finally, the same specimens were calcined for 2 h at 400 °C and measured again by X-ray diffraction.

The abundance of each clay phase in the minerals was obtained from the integrated areas under the d_{001} signal for smectite, illite, dividing each value by an empirical factor established as 1 for illite, 4 for smectite [30]. The textural analysis of the solids was carried out by determining the nitrogen adsorption isotherm at −196 °C, obtained from a 100–200 mg sample in a 3-Flex Micromeritics Sorptometer, over a wide range of relative pressures, previously the samples degassed at 300 °C for 12 h. The BET specific surface areas (S_{BET}) were determined by the multipoint model, using Keii-Rouquerol criteria to find the best linear BET fitting [1]. The external surface (S_{ext}) and the surface corresponding to micropores ($S_{\mu p}$) were calculated using the *t-plot* model. The micropore size distributions were calculated using the method of Horvath and Kawazoe, suitable for the morphology of porous slit type predominant in pillared clays [30].

Hydrogen temperature-programmed reduction analyses (H_2-TPR) were performed using a Micromeritics TPR/TPD 2900 apparatus. About 40 mg sample was heated from room temperature to 900 °C at 10 °C/min under a flow of 60 mL/min of reactive gas (5.0% H_2 in Ar, Air Liquide, Madrid, Spain). Hydrogen consumption was measured with a thermal conductivity detector (TCD), where CuO (Merck 99.99%) was used as the external calibration standard. Based on the measured thermal events, Fe reduction signals were differentiated in terms of the sites present in Al/Fe-PILCs as proposed in advance [13]: fraction of extra-structural FeOx aggregates, fraction of interlayered iron oxide aggregates (iron oxides "decorating" alumina pillars), Fe occupying structural sites of the clay, and finally the Fe forming part of real mixed pillars Al/Fe.

2.4. Catalytic Experiments

The modified materials (C1-P, C2-P, C3-P and BV-P) were evaluated as active solids in hydrogen peroxide-assisted catalytic oxidation of methyl orange (MO). The experiments were carried out at 25.0 ± 1.0 °C and atmospheric pressure (76 kPa) in a 1500 mL Semibatch (Pyrex®, New York, NY, USA) glass reactor equipped with a jacket for temperature control with thermostatic bath and a peristaltic pump to feed the H_2O_2 solution under controlled flow. For each test, the reactor was loaded with 1.0 L of MO solution (Sigma Aldrich, St. Louis, MO, USA, 85%) 0.119 mmol/L and 0.5 g of the solid catalyst to be evaluated (0.05 g of catalyst/L), under constant both air bubbling (about 2 L/min) and mechanical stirring (600 rpm). Addition of 100 mL of the H_2O_2 solution (Panreac, Barcelona, Spain 50%) 51.19 mmol/L (equivalent to exactly the stoichiometric theoretical amount for full mineralization of the azo-dye in the reactor) started after 15 min of stirring and air bubbling (equilibrium period) at flowrate of 2.0 mL/min. The zero time of reaction was the starting point for the addition of the hydrogen peroxide solution into the reactor; from that moment on, 25 mL samples were taken, during a total reaction time of 1 h. The samples were micro-filtered (Millipore, Burlington, MA, USA, 0.45 μm filters) to separate the catalyst previous to analysis. The pH of the solution was adjusted to 3.4 and constantly controlled in this value through the addition of drops of 0.1 mol/L HCl or NaOH [13].

3. Results and Discussion

3.1. Physicochemical Characterization

The chemical compositions of the source minerals (C1, C2, C3 and BV) are shown in Table 1. The SiO_2/Al_2O_3 ratios of materials C2, C3 and BV were much closer to three, indicating the possible presence of smectites [33,34]. Material C1, presented a SiO_2/Al_2O_3 ratio higher than 3.5 and a high content of potassium, which may suggest that this aluminosilicate is mainly constituted of illite. It indirectly shows that C1 presents an important content of collapsed layered-clay phases, since potassium usually lodges strongly in the ditrigonal holes of the tetrahedral layers of the aluminosilicate, hindering any subsequent ion exchange process [31,32]. The presence of iron in all raw minerals (from 7.86 to 19.94% w/w) (Table 1) could be explained because this metal can serve as an isomorphic substitute for both Si and Al within the layered structure. In addition, impurity phases can also be found on the surface of the sheets of the material in the form of extra-structural oxides. This element was present in a higher percentage in the aluminosilicate C3 (19.94% w/w), which may indicate that this mineral exhibits large amount of extra-structural impurities of iron, or that the metal is part of the layers of the material (due to the high content of structural Fe) whose smectitic phase could correspond to nontronite or ferruginous [35].

Table 1. Chemical composition normalized to the content of SiO_2 in the starting aluminosilicates (w/w %) and SiO_2/Al_2O_3 mass ratio in raw minerals, starting clays and pillared materials.

Sample	SiO_2	Al_2O_3	Fe_2O_3	MgO	TiO_2	CaO	K_2O	Na_2O	SiO_2/Al_2O_3
C1	62.94	17.25	7.86	2.23	0.85	3.00	2.03	3.28	3.65
C2	63.30	19.67	8.17	2.64	0.99	2.19	0.99	1.55	3.22
C3	50.77	17.35	19.94	3.08	1.70	2.61	0.17	3.65	2.93
BV	60.25	23.20	9.37	3.16	1.17	0.93	0.80	0.74	2.60
C1-R	55.09	19.33	13.85	2.87	0.68	1.90	1.13	3.25	2.89
C2-R	60.31	21.82	11.02	2.74	1.09	1.05	1.09	0.54	2.76
C3-R	52.04	17.41	17.96	3.25	1.59	2.33	0.20	3.95	2.99
BV-R	57.57	22.02	10.98	2.60	1.24	1.34	0.90	2.88	2.61
C1-P	54.31	27.52	14.65	2.04	0.67	0.14	0.97	0.35	1.97
C2-P	55.33	30.71	13.39	2.07	1.05	0.09	0.83	0.27	1.80
C3-P	49.53	28.03	19.53	2.60	1.54	0.35	0.14	0.17	1.77
BV-P	53.37	33.26	12.58	2.15	1.11	0.08	0.64	0.19	1.60

Regarding the magnesium content, C2 was found to have a lower content (2.64% w/w) compared to BV (3.16% w/w) and C3 (3.08% w/w) (see Table 1), indicating that the smectite phase that may be present in this material would be mostly dioctahedral, whereas BV and C3 could be trioctahedral phyllosilicates. The Na and Ca contents in the starting phyllosilicates are correlated with the exchangeable cations present in their interlayer space. According to the results, the materials with the highest sodium content are C1 (3.28% w/w) and C3 (3.65% w/w), whereas the phyllosilicates C2 and BV have the lowest sodium content (1.55 and 0.74% w/w, respectively), as well as calcium (2.19 and 0.93% w/w, respectively). These characteristics are important if one takes into account that when the interlaminar cation is sodium, the smectites have a greater capacity of swelling in polar solvents. It may conduce to complete dissociation of individual smectite crystals, resulting in a higher degree of dispersion and maximum development of colloidal properties. If Ca or Mg are the predominant exchange cations, their swelling and cationic exchange capacities are much lower [6,21].

The SiO_2/Al_2O_3 ratio also provides information on the amount of quartz that may be present in the starting clays [36]. A high SiO_2/Al_2O_3 ratio indicates a greater amount of quartz in the raw material. Thus, according to Table 1, C1 showed the highest SiO_2/Al_2O_3 ratio with a value of 3.65, followed by C2 (3.22), C3 (2.93), and finally BV (2.60). However, after purification, the SiO_2/Al_2O_3 ratio of these materials decreased as a result of the significant removal of the quartz content; clearly, the most significant decrease was from material C1, suggesting it was the mineral most affected by quartz impurity. In the case of C3, the SiO_2 content indeed became slightly higher after purification, showing that this mineral was the least affected by quartz contamination, but also probably because the loss of iron oxides was more significant.

Semi-quantitative mineralogical analyses were carried out using XRD on monoaxial-oriented films. This mode makes use of the spontaneous stacking property of the laminar clays depending on drying conditions, and allows for films to be arranged in a parallel fashion when set to dry from suspensions diluted at room temperature [31,32]. In layered materials, it is more advantageous than determinations on powdered samples, which given the reduced size of the clay particles, implies that the coherent domain in a single crystal will not give important diffraction maxima and the peaks or signals will become invisible or slightly visible. As XRD quantitative analysis of clay minerals is a very complex task, it should be noted that the one adopted here is a semi-quantitative determination. One of the most complicated variables to control in quantitative approach is the selection of a standard mineral whose diffraction characteristics should be identical to those of every crystalline phase present in the ore-of-interest. Another drawback is that the intrinsic intensity of diffraction of a given mineral also depends on its chemical composition, which is clearly quite variable in this type of natural sample. Powder diffractograms (Figure 2a) of all starting aluminosilicates revealed the presence of illite (I), quartz (Q) and, in the material C2, a very intense signal was observed corresponding to the feldspars

(F) present as impurity in the material. Hence, the semi-quantitative estimation of the mineralogical composition of clay phases present in the raw minerals (Table 2), determined that BV and C3 were the materials with the highest contents of smectite (S), followed by material C2 (60%, 54% and 46%, respectively). Meanwhile, material C1 was basically constituted of non-expandable clay minerals, specifically illite-type (I), and is strongly contaminated by quartz as shown before [37]. These results correlate very well with what was discussed concerning elemental analysis. Then, C1 could be expected to exhibit lower expandability than the rest of the raw materials when intercalated and pillared with the mixed oligomeric Al/Fe system [2,4,21]. In Figure 2b, it is shown that the relative intensity and definition of the reflection corresponding to the basal spacing (001) in smectites significantly increased in the aluminosilicate C2 after carrying out an extensive exchange process with the divalent cation Ca^{2+}, followed by the process of saturation with ethylene glycol in comparison to the raw starting mineral. The same general behavior was observed in the rest of the materials (Supplementary Material, Figure S1). In addition, this reflection in the saturated C2 material slightly shifted to lower 2° theta angle, evidencing the swelling capacity of the smectite present in this aluminosilicate [4]. The smectite phase appears to be well crystallized as it exhibits a peak of strong and acute intensity close to 15.0 Å (Figure 2b), revealing that the predominant exchangeable cation in this aluminosilicate could be calcium and/or probably also magnesium, as suggested by XRF analyses (see Table 1). The signals of the non-expandable phases, commonly found accompanying this type of naturally occurring aluminosilicates, clearly decreased in relative intensity (not shown) as the refining by sedimentation process was carried out; in addition, it generated a remarkable increase in the relative intensity of the signal corresponding to basal spacing in refined materials, especially in BV-R and C2-R solids. It was also reflected in the XRF results (Table 1); as discussed above, both materials showed a decrease in SiO_2 content by about 3.0% with respect to their starting raw minerals (BV and C2).

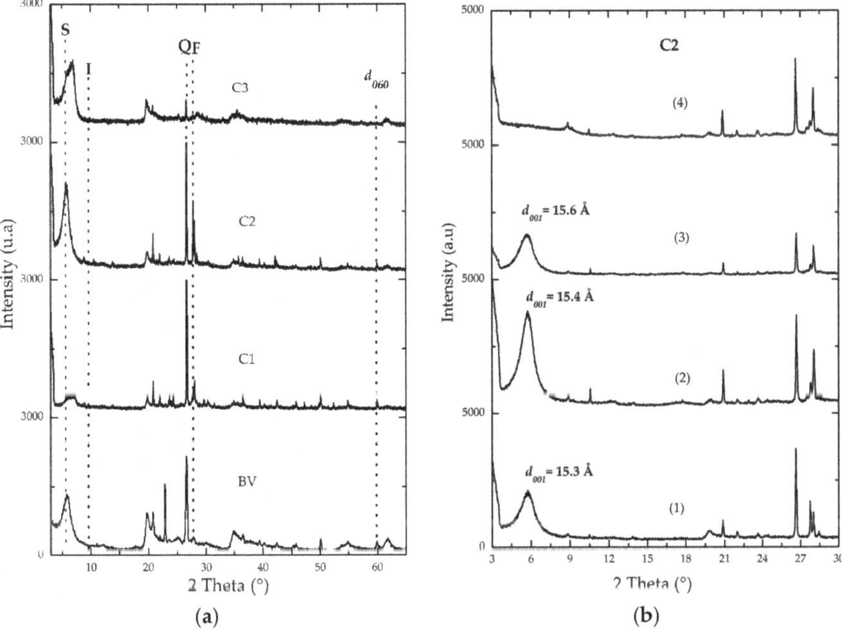

Figure 2. X-ray diffraction patterns of: (**a**) powders of untreated BV, C1, C2 and C3 raw minerals. (S: smectite, I: illite, Q: quartz, F: Feldspars); (**b**) oriented films of C2 clay under following conditions: (1) untreated, (2) Ca^{2+}-homoionized, (3) saturated with ethylene glycol and (4) calcined at 400 °C/2 h.

Table 2. Semi-quantitative mineralogical composition of the clay phases in the raw starting aluminosilicates.

Mineral	Clay Phase	
	Smectite (%)	Illite (%)
BV	60	40
C1	34	66
C2	46	54
C3	54	46

An important increase in the CEC (see Table 3) was observed in the refined materials compared to the raw ones; from C1 to C1-R (40 meq/100 g), from C2 to C2-R (74 meq/100 g) and from BV to BV-R (13 meq/100 g) as a product of the elimination of non-expandable phases such as quartz. However, C3 was the only material that did not show an increase in CEC, but rather a decrease upon refinement (C3-R 10 meq/100 g). It may suggest that at least a part of the content of Fe (19.94% *w/w*, Table 1) in this material, possibly as an impurity, was exchangeable and contributed to the CEC of this mineral, but obviously less in the case of its refined form.

Table 3. Iron incorporated, CEC, compensated fraction of CEC (CC), basal spacing and textural properties of raw, refined, and Al/Fe-pillared aluminosilicates.

Sample	$Fe_{incorporated}$ [a] (Fe_2O_3 w/w %)	CEC [b] (meq/100 g)	% CC [c]	d_{001} [d] (Å)	S_{BET} (m^2/g)	S_{Ext} (m^2/g)	$S_{\mu p}$ (m^2/g)
C1	NA	109	NA	12.8	45	30	15
C2	NA	118	NA	15.3	62	47	15
C3	NA	185	NA	12.7	113	75	37
BV	NA	124	NA	14.7	85	49	37
C1-R	NA	149	NA	16.3	105	78	27
C2-R	NA	192	NA	15.0	105	67	38
C3-R	NA	175	NA	12.6	95	67	28
BV-R	NA	137	NA	15.4	97	64	33
C1-P	0.80	82	45	16.9	144	25	119
C2-P	2.37	78	59	17.4	194	29	165
C3-P	1.57	81	54	17.7	185	48	136
BV-P	1.61	91	50	17.6	139	25	114

NA: Not applicable; [a] Iron content incorporated in the pillared clays; [b] CEC, Cationic Exchange Capacity (dry basis); [c] CC, Compensation of the CEC; [d] Obtained from powder samples.

The d_{060} signal of C1 material showed the trioctahedral nature of its clay phase (59.9°; d = 1.542 Å), whereas dioctahedral for the rest of materials: 62.0° (d = 1.50 Å) for C2, 61.6° (d = 1.504 Å) for C3 and 61.9° (d = 1.504 Å) for BV (Figure 2a). According to several authors, the range of values for this spacing typical in dioctahedral smectites is 1.49 and 1.52 Å, and it oscillates between 1.52 and 1.54 Å for trioctahedral phases [31,32].

The textural properties of the raw materials (Table 3) showed that the one exhibiting the higher S_{BET} was C3 (113 m^2/g); it could be related to its higher Fe content, whose extra-structural fraction may correspond to iron oxides. The extra-structural Fe oxides generally display high external surfaces [38]. In the materials C2 and BV the external surfaces were higher (47 m^2/g and 49 m^2/g) than that of C1 (30 m^2/g), which probably relates to the higher content of clay-like phases present in both materials, whose particle sizes are characterized to be smaller than 2 µm. This is due to the refining step, as non-expandable phases such as quartz were retired, resulting in an increased abundance of clay phases in the refined solids, as well as an increase of the S_{Ext} with respect to the starting minerals, as it can be observed in Table 3. The increase in the specific surface area of the refined material BV-R (97 m^2/g) with respect to its starting raw aluminosilicate (BV, 85 m^2/g) was lower in contrast to the increase observed in C1-R and C2-R refined clays; it correlates with the higher content of smectite present in this material. The refined clay C1-R was the material showing the largest increase in the specific surface area (up to 105 m^2/g) with respect to its crude aluminosilicate (45 m^2/g), which from the point of view of the mineralogical composition is somehow expectable, since it had a higher level of impurities, as discussed above. The starting mineral C3 was the only one whose textural properties did not get enhanced upon refining by particle size; it correlates very well with the above-explained high content of iron oxides displayed by this sample. It is noteworthy that the specific BET surface (S_{BET}) of all the starting raw minerals was mostly represented in external surface, related with the smallest average particle size of the clay fractions, together with their low accessibility into the microporous region (interlayer zone).

As a result of the modification of the starting materials with the mixed Al/Fe oligocationic system (C1-P, C2-P, C3-P and BV-P), all products exhibited an increase in content of both metals (Table 1). However, it should be noted that the C2 material exhibited the highest efficiency incorporating iron (2.37% *w/w* as Fe$_2$O$_3$) and Al (8.89% *w/w* as Al$_2$O$_3$), which happened together with the highest fraction of compensated CEC in the series of minerals (59%; Table 3). It clearly demonstrates that Al/Fe mixed oxides got preferentially stabilized in the interlayer space of the clay, following the targeted cationic exchange mechanism (Table 1). These results are in accordance with those obtained on the widely reported BV-P, reference aluminosilicate [18,26,39]. It got also corroborated as the C2-P pillared material exhibited a decrease of about 90% in its content of the exchangeable cation (Ca^{2+}) relative to its starting, refined material (C2-R) (Table 1). The other materials modified with the mixed Al/Fe system (BV-P, C1-P and C3-P) also showed a decrease in their contents of exchangeable cations (Na$^+$ and Ca^{2+}) with respect to their starting materials, suggesting an exchange and fixation of the intercalating metals in the interlayer space, but not too preferentially as seen for C2-P.

The intercalation of the Keggin-type polyoxocations in C2-P and BV-P materials was also evidenced by the expansion of the basal spacing of these aluminosilicates. It is clearly illustrated through the slight shift of the d_{001} signal to lower 2 theta angles (Figure 3), compared to their corresponding starting, refined materials (C2-R and BV-R). Final basal spacings of both materials were 17.4 Å and 17.6 Å, respectively. These values were only slightly lower than the reported statistical diameter of the Keggin polycation (about 8.9 Å) [2,40,41], which is expected to decrease by heating at high temperatures as in this case. It should be noted that C1-P material exhibited the lowest final basal spacing (16.9 Å, Table 3), probably due to its lower content of expandable phases, therefore leading to lower capacity to intercalate Keggin-type Al/Fe polyoxocations in its interlayer space.

As a result of the pillaring process, the starting clays increased their BET-specific surface areas by approximately 102, 132, 77 and 54 m^2/g for C1-P, C2-P, C3-P and BV-P, respectively (Table 3). Therefore, it can be easily observed that such increase in S_{BET} was predominantly due to formation of microporous surface ($S_{\mu p}$). In addition, it can be inferred that higher textural properties obtained on final pillared material correlated to a great extent with the nature of the starting material previously exhibiting higher CEC, cationic compensation and efficient incorporation of Fe and Al. Therefore, it can be stated that, in general, the CEC of the starting mineral is a fundamental parameter in order to anticipate successful modification via intercalation-pillaring. In this regard, C2 material certainly presented the highest final specific surface area upon Al/Fe-pillaring (C2-P) (194 m^2/g), indicating that it was the aluminosilicate that developed the greatest expansion. C3-P also showed high final specific surface area (185 m^2/g, Table 3) quite comparable to C2-P. However, its external area was larger than the one found in pillared material C2. It means that even after the pillaring step, an important fraction of Fe oxide aggregates persisted on the external surface of the particles, which could be disadvantageous from the catalytic point of view. This type of iron content has been shown in preceding studies to be more easily leached in the CWPO strongly oxidizing environment featuring this reaction [13].

The adsorption–desorption isotherms of the refined and pillared clays are shown in Figure 4. The isotherms of the starting refined materials are type IV (a) according to the IUPAC system (Figure 4 and Figure S2 in Supplementary Material) [42], corresponding to mesoporous materials [43]. The C3-R aluminosilicate presented an adsorption–desorption curve with a higher separation between the branches of adsorption and desorption; it means this material presented high capillary condensation [4].

The isotherms of the pillared aluminosilicates were intermediates between type I, in the range of low relative pressures, and IV, at high p/p^0, according to BDDT classification [44], indicating the presence of both micropores and mesopores. These solids exhibited H3 hysteresis in the IUPAC classification [42], a characteristic behavior of lamellar aggregates of non-homogeneous size and/or shape, whose particles form flexible pores with slit-like morphology [4,45], typical in pillared clays.

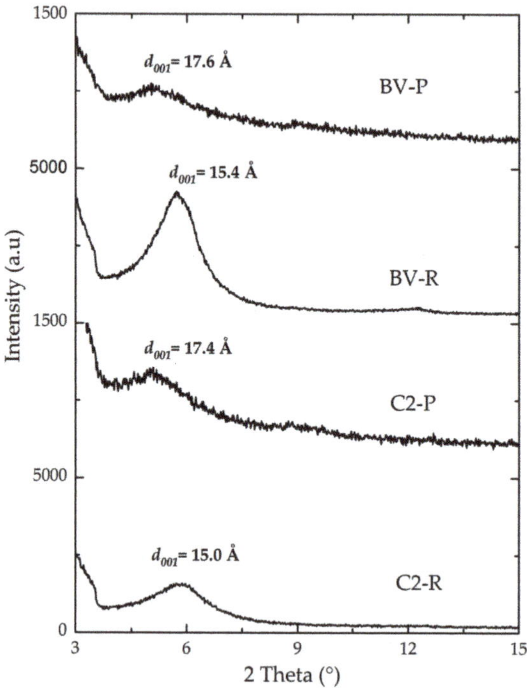

Figure 3. X-ray powder diffraction patterns of refined and pillared materials.

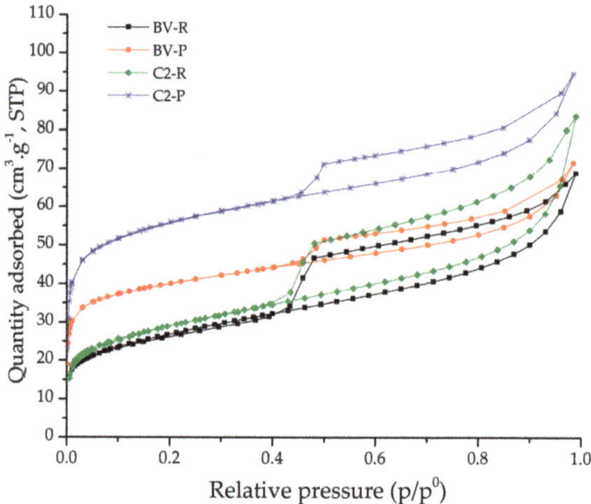

Figure 4. Horvath and Kawazoe micropore width distributions of refined (R) and pillared (P) forms of aluminosilicates C2 and BV.

The Horvath–Kawazoe (HK) distributions of micropore widths (Figure 5) of the pillared materials C2-P and BV-P showed well-defined bimodal behavior: (*i*) a mean pore width around 3.5 Å, which probably corresponds to a fraction of the aluminosilicate interlayered by small, poorly oligomerized Al/Fe species in the intercalating solutions and (*ii*) average pore width centered at 5.0 Å that could be ascribed to the second fraction interlayered with those formerly condensed Keggin-like oligocations, whose statistical diameter (8.9 Å) is obviously expected to decrease after the period of heating at high temperature. Such a last dimension apparently corresponds in relatively poor accord to the expected dimensions for the pores provided by the XRD-measured basal spacings in these two materials (17.4–17.6 Å), after subtraction of the layer thickness featuring the TOT aluminosilicates (around 9.96 Å). However, it is expected to significantly decrease after treatment at high temperature but, in addition, it must be considered that measurements with nitrogen as adsorbate within such a range of very low pressures, corresponding to pore widths below 10 Å, are well-known to deviate in around 2.0 Å towards lower dimensions than actual because of the Van der Waals interactions of nitrogen with such narrow pore walls. It means, the average pore widths truly found in pillared materials C2-P and BV-P were indeed close to 5.5 Å and 7.0 Å, respectively, in higher level of fitting against the theoretical dimensions of the interlayered mixed Keggin-like oligocations. It is noteworthy that a clear difference is observed among the diagrams generated for the pillared forms against those corresponding to the starting, refined forms of both aluminosilicates (C2-R, BV-R); it is obvious that micropore contents measured in the pillared materials were practically absent in the starting, refined materials, whose broad distributions of pore widths were not significantly below 8.0 Å.

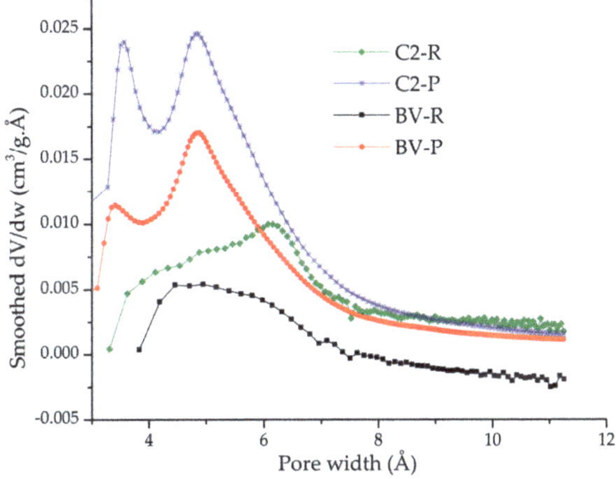

Figure 5. Horvath and Kawazoe micropore width distributions of refined (R) and pillared (P) forms of aluminosilicates C2 and BV.

The H_2-TPR profile (Figure 6) of the starting aluminosilicate C2-R (4307 µmol H_2/g) showed two reduction events centered at around 560 °C and slightly over 800 °C. These can be attributed primarily to the reduction of extra-structural iron (iron oxides, external polluting phases), and secondly to less-reducible structural iron species being part of the aluminosilicate sheets. Meanwhile, C3-R showed three broader, overlapped events of reduction at slightly higher temperatures (450–870 °C) (the curves were fitted into three peaks by the Gaussian method), whose assignments are similar, with the extra signal (peak to 786 °C) at higher temperature probably suggesting the very high Fe content present in this sample (XRF, see Table 1) to be also distributed in various structural sites within the aluminosilicate structural framework (e.g., Fe occupying both octahedral and not only tetrahedral structural holes in the clay layers). A high total hydrogen consumption (13,381 µmol H_2/g) was of course also verified for this material C3-R. The peak deconvolution for diagrams of C3-R and

C3-P materials shows, as expected, a slightly increased total consumption of hydrogen displayed by C3-P against C3-R (see Fe_2O_3 contents, Table 1). In addition, hydrogen consumption of peak 1 in C3-P does not exceed the sum of the first two peaks in thermogram of C3-R. Peak 2 consumption (C3-P) does not exceed sum of peaks 2 and 3 in C3-R. It may suggest that the first peak in C3-R got fully anticipated in temperature in C3-P, but the second peak only partially and then distributed between peaks 1 and 2 in C3-P. Finally, the third peak (broad shoulder) in C3-R got fully anticipated in C3-P second peak. But more interestingly, it is evidently obvious that iron incorporated in C3-P material as a product of the pillaring procedure got widely distributed including extra-structural oxides, being reduced at lower temperature, together with interlayered forms more probably reduced in the second peak of the C3-P diagram. It must be stressed that peak deconvolution of C3-P signal did not consider final consumption (over 680 °C) as an extra peak and then, this consumption is included in the 41.9% displayed in the second peak of the C3-P material. Therefore, hydrogen consumption due to iron interlayered in "truly" mixed Al/Fe-pillars would correspond to such a reduction taking place at the highest temperature.

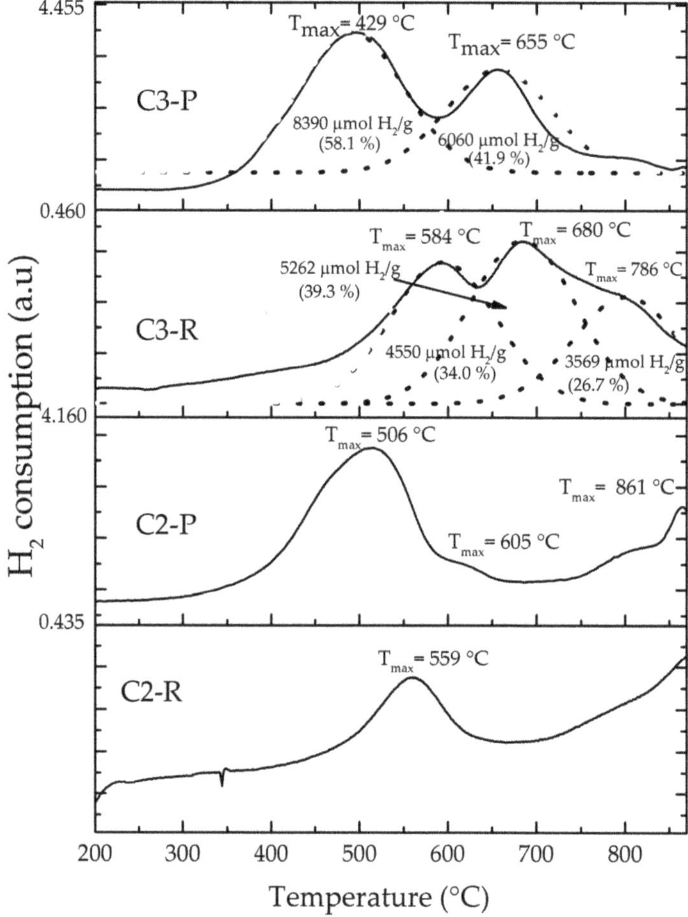

Figure 6. Hydrogen–temperature programmed reduction (H_2–TPR) of refined (R) and pillared forms of clays C2 and C3: Solid lines are experimental curves and dotted lines Gaussian fitted curves of C3-derived materials.

The reduction profiles of all the modified materials (Figure 6 and Figure S3, supplementary material) clearly showed that the first thermal event anticipates its reduction temperature as a result of the pillaring procedure, probably due to increased access to the active sites, making them more easily reduced [16]. However, it should be noted that BV-P and C2-P materials exhibited very similar reduction profiles (Supplementary Material, Figure S3), in which only one instead of two thermal events of reduction was observed at approximately 500 °C corresponding to the aforementioned external oxides. According to bibliographic reports [20], it could be ascribed to FeOOH deposited on the external surface of the clay mineral. This is evidence of the effective pillaring of the C2 aluminosilicate, since the BV-P material, widely reported before, could be used as a reference [45]. In addition, the H_2-TPR diagrams of these materials also showed important hydrogen consumption approaching 900 °C, in good agreement with what was observed elsewhere before [13] for the system Al/Fe-PILC obtained from the same mineral BV, where it was attributed to truly mixed Al/Fe-pillars taking place in the structure. The C2-P material exhibited a shoulder close to 605 °C that, according to earlier interpretation, might correspond to the reduction effect of iron oxide aggregates under strong interaction with Al_2O_3 or the so-called iron "decorating" the alumina pillars [13]. This shoulder may appear overlapped in the C3-P material due to its high content of extra- and structural iron. Besides, this material showed a decrease in the full consumption of hydrogen (12,271 μmol H_2/g) in comparison with its starting material (13,381 μmol H_2/g); it might be due to leaching of a fraction of the extra-structural iron present in the mineral during the pillaring procedure, in good correlation with what was observed in the elemental analyses (Table 1).

3.2. Catalytic Performance of the Al/Fe PILCs

The evolution of decolourization (%) and total organic carbon concentration in the methyl orange solutions throughout the CWPO catalytic reaction using the set of pillared clays as active materials are compared in Figure 7. It can be clearly seen that the C2-P catalyst exhibited the highest catalytic efficiency as a function of both parameters (decolourization about 70%; final TOC content about 10 mg/L, equivalent to around 52.2% of TOC degradation) in only 75 min of reaction at very mild conditions of either temperature (25 °C) of reaction and pH (3.4). These results mainly correlate with the following physicochemical properties of the pillared materials, in this order: (i) the higher amount of incorporated Fe (Table 3), with a high fraction of the transition metal being part of truly mixed Al/Fe pillars according to H_2-TPR diagrams (Figure 6); (ii) the high reached final specific BET surface area (mainly represented as microporous content); (iii) the high starting CEC and its fraction that finally resulted compensated by intercalating polycations (% CC in Table 3). Such a set of physicochemical properties in starting aluminosilicates apparently govern in higher extent the final reactivity of the pillared materials promoting the CWPO catalytic activation of hydrogen peroxide (e.g., internal diffusion of the oxidizing agent towards the active sites present in the catalyst, thus generating a greater efficiency in the production of hydroxyl radicals). In spite of the widely spread criteria around this topic, best performing pillared material did not display higher basal spacing, which suggests that this parameter may vary more widely as a function of the experimental procedure in preparation. Moreover, it was not too different in comparison to the rest of materials studied here or other typical values already reported for Al/Fe-PILCs [4,13]. In addition, it is worth mentioning that previous particle size refining significantly affects the final success of the pillaring process; for instance, C2 material showed only the third higher CEC in raw minerals, but the first in refined forms. It must be stressed that C2-P aluminosilicate presented better catalytic behavior than the BV-P material widely used and reported before in the scientific literature relevant to the field [13,76,77], displaying catalytic performance comparable to those reported in the literature for the CWPO reaction of either methyl orange or orange II in terms of decolourization [13,39,46] using modified clays with the following metal systems: mixed Al/Fe ~80% (75 min of reaction) or MnS ~70% (4 h of reaction). It must be stressed that maximal decolourization reported here was obtained at only 45 min of reaction (plus 30 min of pre-equilibrium period) compared to more than 3 h of reaction [13] or 30 min of reaction (plus 15 min of pre-equillibrium) [39], respectively, reported before. Meanwhile, TOC removal exhibited by Fe-impregnated saponite in decolourization of Orange II azo dye was around 60% (3 h of reaction) [46]. However, it is noteworthy that in some of such cases along with longer times of reaction, higher catalyst loadings were used as compared with this study. Of course, it may display even better maximal values of azo-dye decolourization under optimal conditions of catalyst and peroxide concentrations; it should be noted that a low concentration of catalyst was employed in this study in order to get the highest possible difference in the responses among tested materials. Regarding the effect of the presence of phase impurities such as illite, quartz and feldspar (collapsed or difficult to swell phases) in the starting clay identified in this research, it may influence the physicochemical properties of the starting materials such as the CEC. This means the ability of the aluminosilicate to undergo subsequent ion-exchange processes of the Keggin-like Al/Fe-polyoxocations into interlayer space; of course, it might also indirectly influence the efficiency of incorporation of the active phase, the expansion level of the material and, in turn, final textural properties. However, as far as we know, such a type of impurities have not shown direct influence on the catalytic behavior of the materials (for instance, by transition metals located at structural sites of the phases catalyzing the reaction). In spite of very recent studies claiming the CWPO catalytic activity displayed by naturally occurring minerals [47], they have been mostly iron-rich minerals (namely, hematite, magnetite and ilmenite) exhibiting in addition significant Fe-leaching, pretty longer times and higher temperatures of reaction than those reported here.

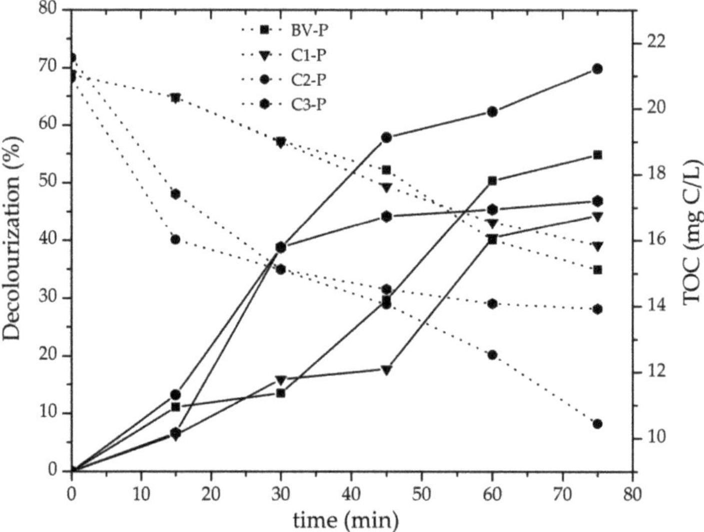

Figure 7. Catalytic behavior of pillared clays in terms of methyl orange decolourization (solid lines) and TOC concentration (dotted lines): Catalyst loading = 0.05 g/L; $[MO]_0$ = 0.119 mmol/L; $[H_2O_2]_{added}$ = 51.2 mmol/L, V_{added} = 100 mL; H_2O_2 stepwise addition = 100 mL/h; pH = 3.4; T = 25 ± 0.1 °C; ambient pressure = 0.76 atm.

Finally, it can be seen in Figure 8 that catalytic performance of C2-P material could not be ascribed to either catalytic response of the Fe content already present in the starting refined mineral, simple adsorption on the surface of the pillared clay or straightforward oxidation by the molecular, non-activated H_2O_2. C2-R material barely reached 20% decolourization and 14.3% of TOC degradation, C2-P material blank (no peroxide added) 6.2% decolourization and 8.0% of TOC degradation, whereas the catalytic system only in the presence of peroxide with no solid catalyst added (peroxide blank) reached less than 20% of bleaching and mineralized only 10.9% of the TOC initial content. It means that the catalytic activity displayed by the C2-P solid was mainly due to the degradation of the azo-dye contaminant by hydroxyl and other radicals generated on the active sites of the modified material, and not to simple adsorption, nor direct attack of the oxidizing agent on the contaminant molecules regarding the effect of the mineralogical composition.

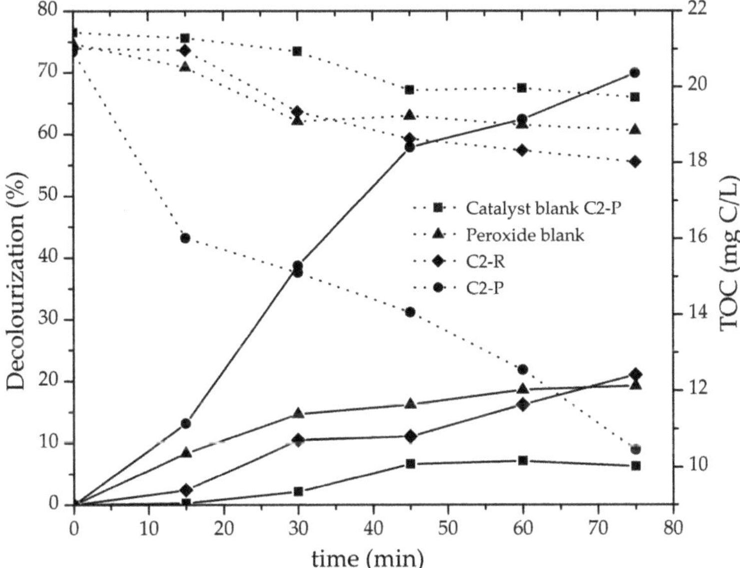

Figure 8. Decolourization (solid lines) and TOC concentration (dotted lines) shown by refined and pillared forms of C2 material in the CWPO degradation of methyl orange: Catalyst loading = 0.05 g/L; $[MO]_0$ = 0.119 mmol/L; $[H_2O_2]_{added}$ = 51.2 mmol/L, V_{added} = 100 mL; H_2O_2 stepwise addition = 100 mL/h; pH = 3.4; T = 25 ± 0.1 °C; ambient pressure = 0.76 atm.

4. Conclusions

In this work, the effects of starting mineral on the physicochemical and catalytic properties of the mixed Al/Fe-pillared clays were studied. It was shown that pillaring procedure on C2-R (starting, previously refined aluminosilicate) favored the best response in the catalytic wet peroxide oxidation of the azo-dye methyl orange in aqueous solution. The highest catalytic performance of this material correlated mainly with following properties of their raw starting and/or refined forms, in this order: the high amount of incorporated Fe with a significant fraction of it being part of truly mixed Al/Fe pillars, high specific BET surface area (mainly represented as microporous content), high starting CEC and its fraction finally compensated by intercalating polycations and basal spacing close to 18 Å. In general, dioctahedral smectites showed better performance in structural modification and catalytic application. In spite of the widely accepted criteria, best performing pillared material did not display higher basal spacing within the study. It suggested that this parameter may closely vary around the targeted space featuring Keggin-like pillared clays, but the textural and catalytic properties deviate seriously as a function of other experimental factors. Low-pressure adsorption isotherms of N_2 treated by the Horvath–Kawazoe (HK) model revealed, for the first time, a bimodal distribution of micropore widths taking place in Al/Fe-PILCs. In summary, pillared aluminosilicate (C2-P) exhibited the best catalytic performance in the CWPO catalytic reaction of methyl orange degradation. This material achieved up to 70% of decolourization and 52% of mineralization of the starting TOC present in solution, under soft reaction conditions (RT, 25 °C, 0.05 g/L of clay catalyst and stoichiometric, no excess of H_2O_2). It was proved that the catalytic performance exhibited by the material was not due to either Fe being present in the starting mineral, simple adsorption of the dye on the catalyst's surface or direct oxidizing effect of the molecular hydrogen peroxide, without previous catalytic effect.

Supplementary Materials: The following are available online at www.mdpi.com/1996-1944/10/12/1364/s1, Figure S1: X-ray diffraction patterns of raw minerals in oriented films: (a) C1, (b) C3 and (c) BV, under following conditions: (1) untreated, (2) Ca^{2+}–homoionized, (3) saturated with ethylene glycol and (4) calcined at 400 °C/2 h, Figure S2: Nitrogen adsorption-desorption isotherms of refined and pillared forms of C1 and C3 clays, Figure S3: Hydrogen–temperature programmed reduction (H_2–TPR) diagrams of refined (R) and pillared forms of clays C1 and BV.

Acknowledgments: The financial support from *Proyecto Agua Potable Nariño* (BPIN 2014000100020), CT&I Fund of the SGR—Colombia is kindly acknowledged. HJM also gratefully thanks the MSc scholarship granted by Departamento de Nariño (BPIN 2013000100092) and managed by CEIBA Foundation, Colombia. M.A.V. and A.G. thank the support from the Spanish Ministry of Economy and Competitiveness (MINECO) and the European Regional Development Fund (FEDER) (projects MAT2013-47811-C2-R and MAT2016-78863-C2-R). A.G. is also grateful for financial support from Santander Bank through the Research Intensification Program.

Author Contributions: L.A.G. and H.J.M. conceived and designed the experiments; H.J.M. performed the experiments; H.J.M., C.B. and L.A.G. analyzed the data; L.A.G. and M.A.V. contributed reagents/materials/ analysis tools; H.J.M., C.B., L.A.G. and A.G. wrote the paper.

Conflicts of Interest: The authors declare no conflicts of interest.

References

1. Rouquerol, J.; Llewellyn, P.; Sing, K. *Adsorption by Clays, Pillared Clays, Zeolites and Aluminophosphates*; Springer: Oxford, UK, 2014; pp. 467–527.

2. Vicente, M.A.; Gil, A.; Bergaya, F. Pillared clays and clay minerals. In *Developments in Clay Science*; Bergaya, F., Lagaly, G., Eds.; Elsevier: Amsterdam, The Netherlands, 2013; Volume 5, pp. 523–557.

3. Lagaly, G. *Colloid Clay Science*; Elsevier: Amsterdam, The Netherlands, 2006; Volume 1, pp. 141–245.

4. Banković, P.; Milutinović-Nikolić, A.; Mojović, Z.; Jović-Jovičić, N.; Perović, M.; Spasojević, V.; Jovanović, D. Synthesis and characterization of bentonites rich in beidellite with incorporated Al or Al–Fe oxide pillars. *Microporous Mesoporous Mater.* **2013**, *165*, 247–256. [CrossRef]

5. Schoonheydt, R.A.; Pinnavaia, T.; Lagaly, G.; Gangas, N. Pillared clays and pillared layered solids. *Pure Appl. Chem.* **1999**, *71*, 2367–2371. [CrossRef]

6. Khankhasaeva, S.; Dambueva, D.V.; Dashinamzhilova, E.; Gil, A.; Vicente, M.A.; Timofeeva, M.N. Fenton degradation of sulfanilamide in the presence of Al,Fe-pillared clay: Catalytic behavior and identification of the intermediates. *J. Hazard. Mater.* **2015**, *293*, 21–29. [CrossRef] [PubMed]

7. Katdare, S.P.; Ramaswamy, V.; Ramaswamy, A.V. Factors affecting the preparation of alumina pillared montmorillonite employing ultrasonics. *Microporous Mesoporous Mater.* **2000**, *37*, 329–336. [CrossRef]

8. Catrinescu, C.; Arsene, D.; Teodosiu, C. Catalytic wet hydrogen peroxide oxidation of para-chlorophenol over Al/Fe pillared clays (AlFePILCs) prepared from different host clays. *Appl. Catal. B Environ.* **2011**, *101*, 451–460. [CrossRef]

9. Cool, P.; Vansant, E.F. Pillared clays: Preparation, characterization and applications. In *Synthesis*; Springer: Berlin/Heidelberg, Germany, 1998; pp. 265–288.

10. Gil, A.; Korili, S.; Trujillano, R.; Vicente, M.A. *Pillared Clays and Related Catalysts*; Springer: New York, NY, USA, 2010.

11. Christidis, G.E. Assessment of industrial clays. In *Developments in Clay Science*; Bergaya, F., Lagaly, G., Eds.; Elsevier: Amsterdam, The Netherlands, 2013; Volume 5, pp. 425–449.

12. Galeano, L.-A.; Vicente, M.Á.; Gil, A. Catalytic degradation of organic pollutants in aqueous streams by mixed Al/M-pillared clays (M = Fe, Cu, Mn). *Catal. Rev.* **2014**, *56*, 239–287. [CrossRef]

13. Galeano, L.A.; Gil, A.; Vicente, M.A. Effect of the atomic active metal ratio in Al/Fe-, Al/Cu- and Al/(Fe–Cu)-intercalating solutions on the physicochemical properties and catalytic activity of pillared clays in the cwpo of methyl orange. *Appl. Catal. B Environ.* **2010**, *100*, 271–281. [CrossRef]

14. Mirzaei, A.; Chen, Z.; Haghighat, F.; Yerushalmi, L. Removal of pharmaceuticals from water by homo/heterogonous fenton-type processes—A review. *Chemosphere* **2017**, *174*, 665–688. [CrossRef] [PubMed]

15. Najjar, W.; Azabou, S.; Sayadi, S.; Ghorbel, A. Catalytic wet peroxide photo-oxidation of phenolic olive oil mill wastewater contaminants. *Appl. Catal. B Environ.* **2007**, *74*, 11–18. [CrossRef]

16. Sanabria, N.; Álvarez, A.; Molina, R.; Moreno, S. Synthesis of pillared bentonite starting from the Al–Fe polymeric precursor in solid state, and its catalytic evaluation in the phenol oxidation reaction. *Catal. Today* **2008**, *133–135*, 530–533. [CrossRef]

17. Molina, C.B.; Zazo, J.A.; Casas, J.A.; Rodriguez, J.J. CWPO of 4-Cp and industrial wastewater with Al-Fe pillared clays. *Water Sci. Technol.* **2010**, *61*, 2161–2168. [CrossRef] [PubMed]

18. Galeano, L.A.; Bravo, P.F.; Luna, C.D.; Vicente, M.Á.; Gil, A. Removal of natural organic matter for drinking water production by Al/Fe-PILC-catalyzed wet peroxide oxidation: Effect of the catalyst preparation from concentrated precursors. *Appl. Catal. B Environ.* **2012**, *111–112*, 527–535. [CrossRef]

19. Banković, P.; Milutinović-Nikolić, A.; Mojović, Z.; Jović-Jovičić, N.; Žunić, M.; Dondur, V.; Jovanović, D. Al,Fe-pillared clays in catalytic decolorization of aqueous tartrazine solutions. *Appl. Clay Sci.* **2012**, *58*, 73–78. [CrossRef]

20. Gao, Y.; Li, W.; Sun, H.; Zheng, Z.; Cui, X.; Wang, H.; Meng, F. A facile in situ pillaring method—The synthesis of Al-pillared montmorillonite. *Appl. Clay Sci.* **2014**, *88–89*, 228–232. [CrossRef]

21. Sun, L.; Ling, C.Y.; Lavikainen, L.P.; Hirvi, J.T.; Kasa, S.; Pakkanen, T.A. Influence of layer charge and charge location on the swelling pressure of dioctahedral smectites. *Chem. Phys.* **2016**, *473*, 40–45. [CrossRef]

22. McCabe, R.W.; Adams, J.M. Clay minerals as catalysts. In *Developments in Clay Science*; Bergaya, F., Lagaly, G., Eds.; Elsevier: Amsterdam, The Netherlands, 2013; Volume 5, pp. 491–538.

23. Zeng, L.; Wang, S.; Peng, X.; Geng, J.; Chen, C.; Li, M. Al–Fe PILC preparation, characterization and its potential adsorption capacity for aflatoxin B1. *Appl. Clay Sci.* **2013**, *83–84*, 231–237. [CrossRef]

24. Catrinescu, C.; Arsene, D.; Apopei, P.; Teodosiu, C. Degradation of 4-chlorophenol from wastewater through heterogeneous fenton and photo-fenton process, catalyzed by Al–Fe PILC. *Appl. Clay Sci.* **2012**, *58*, 96–101. [CrossRef]

25. Li, K.; Lei, J.; Yuan, G.; Weerachanchai, P.; Wang, J.-Y.; Zhao, J.; Yang, Y. Fe-, Ti-, Zr- and Al-pillared clays for efficient catalytic pyrolysis of mixed plastics. *Chem. Eng. J.* **2017**, *317*, 800–809. [CrossRef]

26. Garcia, A.M.; Moreno, V.; Delgado, S.X.; Ramírez, A.E.; Vargas, L.A.; Vicente, M.Á.; Gil, A.; Galeano, L.A. Encapsulation of SALEN- and SALHD-Mn(III) complexes in an Al-pillared clay for bicarbonate-assisted catalytic epoxidation of cyclohexene. *J. Mol. Catal. A Chem.* **2016**, *416*, 10–19. [CrossRef]

27. Sanabria, N.R.; Ávila, P.; Yates, M.; Rasmussen, S.B.; Molina, R.; Moreno, S. Mechanical and textural properties of extruded materials manufactured with AlFe and AlCeFe pillared bentonites. *Appl. Clay Sci.* **2010**, *47*, 283–289. [CrossRef]

28. Urruchurto, C.M.; Carriazo, J.G.; Osorio, C.; Moreno, S.; Molina, R.A. Spray-drying for the preparation of Al–Co–Cu pillared clays: A comparison with the conventional hot-drying method. *Powder Technol.* **2013**, *239*, 451–457. [CrossRef]

29. Barrault, J.; Abdellaoui, M.; Bouchoule, C.; Majesté, A.; Tatibouët, J.M.; Louloudi, A.; Papayannakos, N.; Gangas, N.H. Catalytic wet peroxide oxidation over mixed (Al–Fe) pillared clays. *Appl. Catal. B Environ.* **2000**, *27*, L225–L230. [CrossRef]

30. Bergaya, F.; Hassoun, N.; Barrault, J.; Gatineau, L. Pillaring of synthetic hectorite by mixed [Al13-Xfex] pillars. *Clay Miner.* **1999**, *28*, 109–122. [CrossRef]

31. Thorez, J. *Practical Identification of Clay Minerals: A Handbook for Teachers and Students in Clay Mineralogy*, 1st ed.; Lelotte: Berkeley, CA, USA, 1976.

32. Moore, D.; Reynolds, R., Jr. *X-ray Diffraction and the Identification and Analysis of Clay Minerals*, 2nd ed.; Oxford University Press: Oxford, UK, 1997; p. 373.

33. Tsozué, D.; Nzeugang, A.N.; Mache, J.R.; Loweh, S.; Fagel, N. Mineralogical, physico-chemical and technological characterization of clays from maroua (Far-North, Cameroon) for use in ceramic bricks production. *J. Build. Eng.* **2017**, *11*, 17–24. [CrossRef]

34. Nguetnkam, J.P.; Kamga, R.; Villiéras, F.; Ekodeck, G.E.; Yvon, J. Altération différentielle du granite en zone tropicale. Exemple de deux séquences étudiées au cameroun (Afrique centrale). *C. R. Geosci.* **2008**, *340*, 451–461. [CrossRef]

35. Franco, F.; Pozo, M.; Cecilia, J.A.; Benítez-Guerrero, M.; Lorente, M. Effectiveness of microwave assisted acid treatment on dioctahedral and trioctahedral smectites. The influence of octahedral composition. *Appl. Clay Sci.* **2016**, *120*, 70–80. [CrossRef]

36. Chin, C.L.; Ahmad, Z.A.; Sow, S.S. Relationship between the thermal behaviour of the clays and their mineralogical and chemical composition: Example of Ipoh, Kuala Rompin and Mersing (Malaysia). *Appl. Clay Sci.* **2017**, *143*, 327–335. [CrossRef]

37. Brown, G.; Newman, A.C.D.; Rayner, J.H.; Weir, A.H. The structures and chemistry of soil clay minerals. In *The Chemistry of Soil Constituents*; Greenland, D.J., Hayes, M.H.B., Eds.; John Wiley & Sons: New York, NY, USA, 1978; pp. 29–178.

38. Eusterhues, K.; Rumpel, C.; Kogel-Knabner, I. Organo-mineral associations in sandy acid forest soils: Importance of specific surface area, iron oxides and micropores. *Eur. J. Soil Sci.* **2005**, *56*, 753–763. [CrossRef]

39. Galeano, L.A.; Gil, A.; Vicente, M.A. Strategies for immobilization of manganese on expanded natural clays: Catalytic activity in the cwpo of methyl orange. *Appl. Catal. B Environ.* **2011**, *104*, 252–260. [CrossRef]

40. Zhu, R.; Chen, Q.; Zhou, Q.; Xi, Y.; Zhu, J.; He, H. Adsorbents based on montmorillonite for contaminant removal from water: A review. *Appl. Clay Sci.* **2016**, *123*, 239–258. [CrossRef]

41. Gonzalez, B.; Perez, A.H.; Trujillano, R.; Gil, A.; Vicente, M.A. Microwave-assisted pillaring of a montmorillonite with Al-polycations in concentrated media. *Materials* **2017**, *10*, 886. [CrossRef] [PubMed]

42. Thommes, M.; Kaneko, K.; Neimark, A.V.; Olivier, J.P.; Rodriguez-Reinoso, F.; Rouquerol, J.; Sing, K.S.W. Physisorption of gases, with special reference to the evaluation of surface area and pore size distribution (IUPAC technical report). *Pure Appl. Chem.* **2015**, *87*, 1051–1069. [CrossRef]

43. Kooli, F. Pillared montmorillontes from unusual antiperspirant aqueous solutions: Characterization and catalytic tests. *Microporous Mesoporous Mater.* **2013**, *167*, 228–236. [CrossRef]

44. Sing, K.S.W.; Everett, D.H.; Haul, R.A.W.; Moscou, L.; Pierotti, R.A.; Rouquérol, J.; Siemieniewska, T. Reporting physisorption data for gas/solid systems with special reference to the determination of surface area and porosity. *Pure Appl. Chem.* **1985**, *57*, 603–619. [CrossRef]

45. Carriazo, J.G. Influence of iron removal on the synthesis of pillared clays: A surface study by nitrogen adsorption, XRD and EPR. *Appl. Clay Sci.* **2012**, *67–68*, 99–105. [CrossRef]

46. Herney-Ramirez, J.; Lampinen, M.; Vicente, M.A.; Costa, C.A.; Madeira, L.M. Experimental design to optimize the oxidation of orange II dye solution using a clay-based fenton-like catalyst. *Ind. Eng. Chem. Res.* **2008**, *47*, 284–294. [CrossRef]

47. Munoz, M.; Domínguez, P.; de Pedro, Z.M.; Casas, J.A.; Rodriguez, J.J. Naturally-occurring iron minerals as inexpensive catalysts for CWPO. *Appl. Catal. B Environ.* **2017**, *203*, 166–173. [CrossRef]

 *materials* 

Article

Removal of Ciprofloxacin from Aqueous Solutions Using Pillared Clays

Maria Eugenia Roca Jalil [1],*, Miria Baschini [1] and Karim Sapag [2]

[1] Grupo de Estudios en Materiales Adsorbentes, Instituto de Investigación y Desarrollo en Ingeniería de Procesos, Biotecnología y Energías Alternativas-CONICET, Universidad Nacional del Comahue, Buenos Aires, 1400 8300 Neuquén, Argentina; miria.baschini@fain.uncoma.edu.ar

[2] Laboratorio de Sólidos Porosos, Instituto de Física Aplicada—CONICET, Universidad Nacional de San Luis, Ejército de los Andes 950, Bloque II, 2do piso, CP 5700 San Luis, Argentina; sapag@unsl.edu.ar

* Correspondence: eugenia.rocajalil@fain.uncoma.edu.ar; Tel.: +54-299-449-0300 (ext. 688)

Received: 10 October 2017; Accepted: 20 November 2017; Published: 23 November 2017

Abstract: Emerging contaminants in the environment have caused enormous concern in the last few decades, and among them, antibiotics have received special attention. On the other hand, adsorption has shown to be a useful, low-cost, and eco-friendly method for the removal of this type of contaminants from water. This work is focused on the study of ciprofloxacin (CPX) removal from water by adsorption on pillared clays (PILC) under basic pH conditions, where CPX is in its anionic form (CPX^-). Four different materials were synthetized, characterized, and studied as adsorbents of CPX (Al-, Fe , Si , and Zr PILC). The highest CPX adsorption capacities of 100.6 and 122.1 mg g^{-1} were obtained for the Si- and Fe-PILC (respectively), and can be related to the porous structure of the PILCs. The suggested adsorption mechanism involves inner-sphere complexes formation as well as van der Waals interactions between CPX^- and the available adsorption sites on the PILC surfaces.

Keywords: ciprofloxacin; adsorption; pillared clays

---

1. Introduction

The presence of antibiotics in wastewater and surface water has been widely reported [1–3], and is becoming a growing concern due to its toxicological effect on aquatic species as well as the resistance that they can induce on some bacterial strains even at low concentrations. These compounds reach the environment as consequence of different activities, like veterinary medicine or agriculture but, mostly, due to their use in human medicine and the inefficiency of wastewater treatments to remove this kind of contaminants, which are not biodegraded [1–6]. Consequently, different methods have been studied to eliminate antibiotics from water, such as advanced oxidation processes, nanofiltration, reverse osmosis, and photo and electrochemical degradation. The downsides of these approaches go from high maintenance costs, to complicated procedures, or secondary pollution [1–6].

From the available methods, the adsorption process has shown to be the most effective and inexpensive. It also involves an easier design and operation than other techniques. Moreover, there is a huge variety of adsorbents with different properties and nature, such as carbonous materials [7–12], mesoporous silica [13], hydrous oxides [14], and mineral clays [15–19], which have been evaluated for these new organic pollutants. Although activated carbons have been extensively used for the removal of organic compounds from water, mineral clays have acquired attention because they are effective adsorbents, as well as low-cost, widely distributed, and eco friendly materials. In particular, bentonites have been greatly studied for removing emergent contaminants as amoxicillin [20], diclofenac potassium [21], tetracycline [22,23], cephalexin [24], and ciprofloxacin [15,17–19] from water. However, the bentonites bring limitations in their separation from the aqueous media due to their behavior in water suspensions. Pillared clays (PILC) are micro-mesoporous materials that are synthetized from bentonites characterized by a large specific surface area, permanent porosity, and higher hydrophobicity than the one shown by the raw material [25]. Although PILC were obtained as alternative catalysts to zeolites and have been widely studied both as catalysts and as catalyst supports, these materials have also proven to be effective adsorbents of diverse organic and inorganic pollutants [26–34].

Taking this into account, four pillared clays with different oligocations (Al, Fe, Si, and Zr) were synthetized and characterized. The obtained PILC were evaluated as adsorbents of ciprofloxacin from basic aqueous media in order to study the relationship between their structural and textual properties and their removal capacity.

2. Materials and Methods

2.1. Synthesis and Characterization of Pillared Clays

The natural clay (NC) used as raw material in the present work is a bentonite that is obtained from the Pellegrini lake in the province of Rio Negro, Argentina. Four pillared clays with different pillaring agents were synthetized from this NC, described in detail in a previous work [17].

The silica pillared clay (Si-PILC) was prepared following the methodology described by Han et al. [35], with some modifications. The silica sol solution that was used as pillaring agent was obtained by mixing tetraethyl orthosilicate (TEOS: $Si(OEt)_4$, Merck > 99%), 2 M HCl (Cicarelli, 36.5–38%) and ethanol in a molar ratio of 1:0.1:1. The resulting solution was then aged at room temperature for 2 h and was mixed with a 0.25 M ferric nitrate ($Fe(NO_3)_3$ $9H_2O$) solution in a molar ratio of Si/Fe 10:1. This mixture was titrated with 0.2 M NaOH (Anedra, 98%) solution up to a pH of 2.7. Then, the pillaring agent obtained was added drop wise to a NC suspension of 1 wt % of deionized water in a molar ratio of Si/Fe/CIC 50:5:1, respectively. During the cation exchange, the mixture was stirred 3 h at 60 °C and the solid was separated by centrifugation at 3500 rpm for 15 min by Sorvall RC 5C centrifuge (Kendro Laboratory Product, Newtown, CT, USA). This solid (the exchanged NC), was first washed with a solution of ethanol/water 50% v/v to remove excess oligocation solution, and was then dispersed in 0.2 M HCl solution under stirring during 3 h to leach out the iron species from the mixed sol particles. This last step was carried out four more times. Finally, the solid material was washed with deionized water several times and then dried to finally obtain the Si-pillared clay precursor.

The iron pillared clay (Fe-PILC) was synthesized according to the procedure proposed by Yamanaka et al. [36], where the trinuclear acetate-hydroxo iron (III) nitrate ($[Fe_3O(OCOCH_3)_6CH_3COOH(H_2O)_2]NO_3$) is the pillaring agent. The latter was prepared mixing a ferric nitrate ($Fe(NO_3)_3$ $9H_2O$, Fluka, 97%) solution with ethanol and adding drop by drop acetic anhydride in a molar ratio of 1:4.3:7.4, respectively. The solution was further kept in an ice bath for cooling and the resulting precipitate was separated by filtration. Then, a 0.04 M aqueous solution of trinuclear acetate complex was prepared and added to a 1 wt % suspension of the NC in deionized water. The mixture was stirred for 3 h and the resulting suspension was then filtered, washed with deionized water, and dried, to finally obtain the Fe-pillared clay precursor.

The aluminum pillared clay (Al-PILC) was prepared following the methodology described in previous work [32]. The pillaring agent was synthesized by basic hydrolysis of $AlCl_3 \cdot 6H_2O$ (Anedra, Lieshout, NL, 99.5%) solution with NaOH solution. The basicity relationship was of $OH^-/Al^{3+} = 2$ and the mixture was initially maintained under stirring at 60 °C, and then aged under stirring for 12 h at room temperature. The pillaring agent obtained was added drop wise to a 3 wt % dispersion of NC in deionized water and it was stirred for 1 h. The resulting solid was washed using dialysis membranes and was dried to obtain the Al-PILC precursor.

The zirconium pillared clay (Zr-PILC) was synthesized using the methodology proposed by Farfan Torres et al. [37] with modifications. The pillaring agent used was a $ZrOCl_2$ $8H_2O$ (Merck, Kenilworth, NJ, USA, 99.9%) solution initially adjusted at a pH of 1.9 using a NaOH solution. The resulting solution was added drop by drop to a 1 wt % suspension of NC in deionized water under stirring at 40 °C.

The mixture was then stirred for 2 h, filtered, washed with deionized water, and dried for obtaining the Zr-PILC precursor.

Finally, all of the precursors of the PILC were calcined at 500 °C for 1 h in order to obtain the Si-PILC, Fe-PILC, Al-PILC, and Zr-PILC samples.

The structural properties for all of the materials were analyzed by X-ray Diffraction (XRD) using a RIGAKU Geigerflex X-ray diffractometer (Rigaku, Austin, TX, USA with CuKα radiation at 20 mA and 40 kV. The scans were recorded between 2° and 70° (2θ), with a step size of 0.02° and a scanning speed of 2° min^{-1}. The textural properties were studied by nitrogen adsorption-desorption isotherms at -196 °C. These measurements were carried out using an Autosorb 1 MP and iQ (Quantachrome, Boynton Beach, FL, USA). All of the samples were previously degassed for 12 h up to a residual pressure lower than 0.5 Pa at 200 °C. Textural properties were obtained from these isotherms by different methods. The specific surface area (S_{BET}) was assessed by the Brunauer, Emmet and Teller (BET) method, using the Rouquerol's criteria [38]. The micropore volumes ($V_{\mu p}$) were calculated with the *α-plot* method using the corresponding sample calcined at 1000 °C as reference material [39]. The total pore volume (V_T) was obtained using the Gurvich rule (at 0.97 of relative pressure) [38]. Pore size distributions (PSD) were obtained by the Horvarth-Kawazoe method, when considering the adsorption branch and that the PILC have slit shape pores within the interlayer region.

2.2. Adsorptive: Ciprofloxacin

The ciprofloxacin (CPX) is an antibiotic from the fluoroquinolones group, which is widely used in human health. The CPX is a crystalline solid with a molecular weight of 331.4 g mol^{-1}. The dimensions of the (CPX) molecule are 1.35 nm $\times$ 0.3 nm $\times$ 0.74 nm and a scheme of its chemical structure is shown in Figure 1 [8].

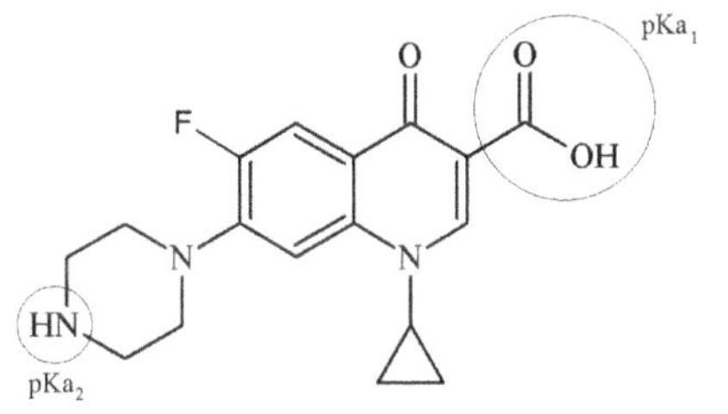

Figure 1. Ciprofloxacin structure.

The presence of protonable groups in the CPX structure generates two pKa values, and, in consequence, three possible species in solution. The pKa values for the CPX are 5.90 ± 0.15 (pKa$_1$) and 8.89 ± 0.11 (pKa$_2$), which correspond to the carboxylic acid group and the amine group in the piperazine moiety, respectively (see Figure 1). The protonation-deprotonation reactions that take place at different pH values of the media affect the species that are present in the solution, as well as their solubility. Therefore, the CPX can be found as cation (CPX$^+$), zwiterion (CPX$^\pm$), or anion (CPX$^-$) under different pH values where the zwiterion form shows the lowest solubility [17].

In this work, the ciprofloxacin hydrochloride was used as adsorptive; it was acquired from Romikim S.A (CHEMO Argentina, Buenos Aires, Argentine) and had 99.3% of purity.

2.3. CPX Adsorption Studies

The adsorption experiments were conducted by mixing 0.02 g of adsorbent with 8 mL of CPX solutions from 18 to 230 mg·L^{-1} in tubes of 10 mL and further stirring at 20 °C up to beyond the equilibrium time. The values were chosen according to previous results [17]. In all of the tests, the tubes were wrapped in aluminum foils to prevent light-induced decomposition. After the adsorption, the solutions were separated from the adsorbent using a Sorvall RC 5C centrifuge at 8000 rpm for

20 min. The CPX equilibrium concentrations in the resultant supernatant were measured with a T60 UV-vis spectrophotometer (PG Instruments Lmited, Leicester, UK) at the λ_{max} corresponding to the pH value, from the previously determined calibration curve. The absorption spectra were obtained for every point of each isotherm. In all of cases the spectra observed were the same as those of CPX before clay contact, evidencing that CPX is not degraded after contact with PILC materials. All of the samples were measured in duplicate and the average value was used. The amount of CPX adsorbed on the clay mineral (q) was calculated from the initial and equilibrium CPX concentrations, according to an Equation (1):

$$q = \frac{V\left(C_i - C_{eq}\right)}{w} \tag{1}$$

where V is the CPX solution volume (L), C_i is the initial CPX concentration (mg L^{-1}), C_{eq} is the equilibrium CPX concentration (mg L^{-1}), and w is the mass of clay (g).

2.4. Modelling Methods

Langmuir, Freundlich, and Sips isotherms models were fitted to the CPX adsorption equilibrium data [40]. The Langmuir model assumes a monolayer adsorption on a surface with a finite number of identical sites, which are energetically equivalent and where there are no interactions among the adsorbed molecules (homogeneous surface). The mathematical expression of the Langmuir model is shown in Equation (2):

$$q = \frac{q_m k C_{eq}}{1 + k C_{eq}} \tag{2}$$

where q_m is the maximum adsorbed concentration within a monolayer of adsorbate (mg g^{-1}) and k (L mg^{-1}) is the Langmuir adsorption equilibrium constant, which is related to the adsorption energy.

The Freundlich equation is an empirical method that has been widely applied to adsorption on heterogeneous surfaces. This model uses a multi-site adsorption isotherm and its mathematical expression is defined in Equation (3).

$$q = k_F C_{eq}^{1/n} \tag{3}$$

where k_F (mg g^{-1}(L mg^{-1})^n) and n (dimensionless) are the Freundlich characteristic constants, indicating the adsorption capacity and adsorption intensity, respectively.

The Sips equation is a combination of the Langmuir and Freundlich equations. It is an empirical equation that assumes a heterogeneous surface with a number of active sites that interact with the adsorbate molecule, without adsorbate-adsorbate interactions. The mathematical expression is shown in Equation (4).

$$q = q_m \frac{\left(bC_{eq}\right)^{1/n}}{1 + \left(bC_{eq}\right)^{1/n}} \tag{4}$$

where q_m and C_{eq} have the same meanings as above, b is a parameter related to the affinity of the adsorbate towards the surface, and n is a parameter that represents the heterogeneity of the system.

The Scatchard model was applied to the adsorption data to gather complementary information about the adsorption phenomena. This method involves the transformation of the data from the isotherm to obtain a plot of q/C_{eq} versus q (where q has the same meaning indicated above). The resulting plot is called a Scatchard plot and if it is a straight line, it suggests that the adsorption takes place in the same type of sites. On the other hand, if the plot is a non-linear curve, its shape can be related to nonspecific or multi-type interactions between the adsorbate and adsorbent surface. Concave curves are related to negative cooperative phenomena or to the presence of heterogeneity sites for the adsorption. In turn, convex curves are associated to positive cooperative phenomena where the first adsorption occurs with low affinity and that the adsorbate becomes a possible site for the subsequent adsorption. Additionally, any deviation from linearity in the Scatchard plot (taking R^2 values) could be considered as an indication of the presence of nonspecific or multi-type interactions of the adsorbate molecules towards the surface sites [41–44].

3. Results and Discussion

3.1. Adsorbents Characterization

XRD patterns obtained for the PILC materials were evaluated in contrast to those obtained for the NC and the basal distances (d_{001}). The d_{001} value obtained for NC was 1.26 nm, which is typical of natural sodic montmorillonites, whereas Al- and Fe-PILC showed values of 1.85 and 1.60 nm, respectively. These increases in the basal distances for the PILC could be attributed to the presence of aluminum and ferric oxide species within the NC interlayer. Additionally, the diffractograms did not show any other structural changes in the PILC in relation to the starting material. In the case of the Si- and Zr-PILC, the patterns obtained do not show any defined peaks. The low resolution for these materials could be due to a higher heterogeneity in their structures in comparison to the NC. This heterogeneity was possibly caused by a stronger acid media used in their synthesis, which may have affected the usual order of the raw material layers [45,46].

Nitrogen adsorption-desorption isotherms at $-196\ ^{\circ}C$ of the adsorbents are shown in Figure 2, where the amount adsorbed is expressed in cubic centimeters at STP (standard temperature and pressure) per gram of material ($V_{ads}\ cm^3$, STP g^{-1}) plotted against the equilibrium relative pressure (p/p^0). The shape of adsorption-desorption isotherms are grouped into six types and can be related to particular pore structures according to the IUPAC (International Union of Pure and Applied Chemistry) classification [38,47]. Taking this in account, the natural clay isotherm can be classified as type IIb isotherm with an H3 hysteresis loop. This kind of isotherm is associated to mesoporous materials with aggregates of plate-like particles, like the montmorillonites. The isotherms that are obtained for the pillared clays can be classified as a combination between type I, at low relative pressures, and type IIb due to the adsorption behavior in the mono-multilayer region (from 0.05 to ca. 0.8 in relative pressures). The high amount adsorbed at low relative pressures is related to the microporosity that is generated as a result of the pillaring process. In addition, the presence of mesoporosity in PILC is evidenced by the presence of the type H4 hysteresis loops that are associated to slit-like pores, as well as to the pores that are generated within the interlayer of the clay minerals. Among the pillared clays, the Si-PILC showed the highest adsorption values at low relative pressure, suggesting that this material developed a greater microporosity. Regarding Al-, Fe-, and Zr-PILC, the adsorption in the micropores region was lower than the one that was obtained for the Si-PILC, but was still higher than the one obtained for the natural clay, suggesting a successful pillaring process.

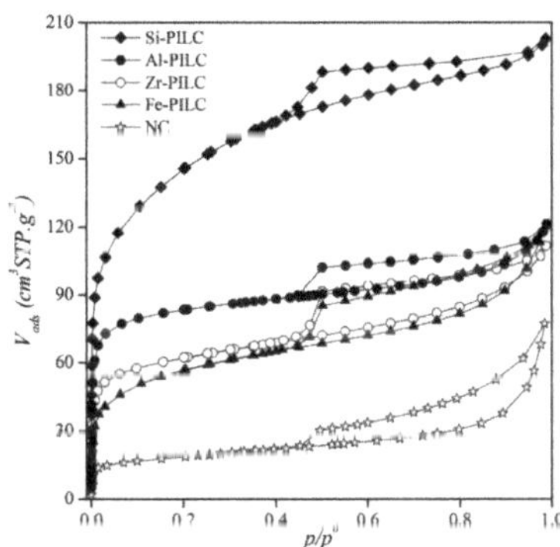

Figure 2. N_2 adsorption-desorption isotherms at 77 K for natural and pillared clay minerals.

The textural properties obtained for all of the materials from nitrogen adsorption-desorption isotherms are summarized in Table 1.

Table 1. Textural properties data and d_{001} obtained for natural and pillared clays.

Materials	S_{BET} (m^2 g^{-1})	V_T (cm^3 g^{-1})	$V_{\mu p}$ (cm^3 g^{-1})
Natural Clay	67	0.10	0.01
Al-PILC	322	0.18	0.12
Si-PILC	519	0.31	0.19
Zr-PILC	231	0.16	0.07
Fe-PILC	206	0.17	0.07

Data showed an increase in the specific surface areas (S_{BET}) for all PILC in comparison with the raw material, and, due to the microporosity accomplished by the pillaring process. The highest and lowest S_{BET} values were eight and three times the NC value for Si- and Fe-PILC, respectively. The order of increase of S_{BET} values was the same as the micropores volume ($V_{\mu p}$) for all of the materials, where the values obtained for Si- and Al-PILC were around two times higher than the ones that were obtained for other PILC. This suggests a higher density of pillars for these materials than for the other PILC, and it could be related to the synthesis method. The percentages of microporosity were between 41% and 66% for Zr- and Al-PILC, respectively. Si- and Al-PILC were the materials exhibiting the highest microporosity percentages, suggesting pillaring agents with more uniform structures and greater pillars density within the interlayer. The V_T values supported the behavior of the isotherms at high relative pressures. The highest value was obtained for the Si-PILC, suggesting a higher porosity in this sample as compared to the other PILC.

The pore size distributions (PSD) for the PILC are shown in Figure 3. All of them were studied in the microporous region (pores size below 2 nm) to comparatively follow the development of microporosity. Al-PILC has micropore sizes of between 0.4 and 2 nm and are higher than the one observed for the NC. However, Si- and Zr-PILC showed PSD with more defined micropore sizes around 0.6 and 0.8 nm, respectively. These results could suggest that a high density of pillars was caused during the pillaring process in these materials. The PSD obtained for the Al-PILC shows that this material has micropore sizes of around 0.9 nm, which are the characteristic dimensions of the kegging cation that is used as pillaring agent [32]. The Fe-PILC showed fewer micropores than the other PILC and the PSD shape suggested micropores with different sizes in its interlayer. All of the results showed an agreement with the textural properties.

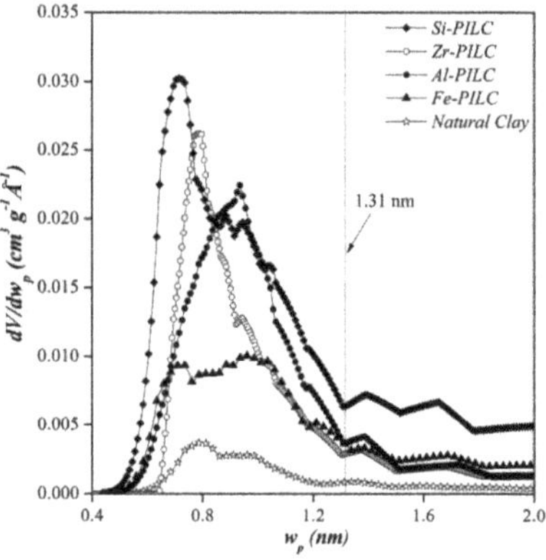

Figure 3. Pore side distribution of natural and pillared clay minerals where *V* is adsorbed volume and w_p is the pore size.

3.2. Effect of pH Media on Adsorption

The studies of CPX adsorption on NC and PILC at different pH values were assessed under the above mentioned conditions, with an initial fixed CPX concentration of 110 mg L^{-1}. The pH of the initial solution was adjusted to values between 3 and 12 using HCl or NaOH solutions. This value was chosen based on the CPX solubility that was measured in previous work [17]. Adsorption capacities of CPX on NC and PILC at different pH values are shown in Figure 4.

The results obtained for NC exhibited highest CPX adsorption at low pH values, decreasing with the pH increase from 7.5, which can be explained by the relationship between the surface charge of NC and the CPX species that are present. At low pH values, the species of CPX that are present are in their cationic form, favoring the adsorption on negative charged NC surface by cation exchange of CPX$^+$ for the natural cation within the montmorillonite interlayer. This is the adsorption mechanism that is proposed and widely reported for adsorption of cations on montmorillonites [17,18,22,48]. The decrease in CPX adsorption after a pH of 7.5 can be explained by the presence of the zwiterionic and anionic forms of CPX. In these cases, the presence of a negative charge in the CPX structure results in repulsive interactions with the mineral negative surface, resulting in other adsorption mechanisms. The highest amount adsorbed for the NC was obtained at a pH of 6, probably due to the competitive adsorption of the H$^+$ against the CPX$^+$ species towards the same sites on the clay surface [17].

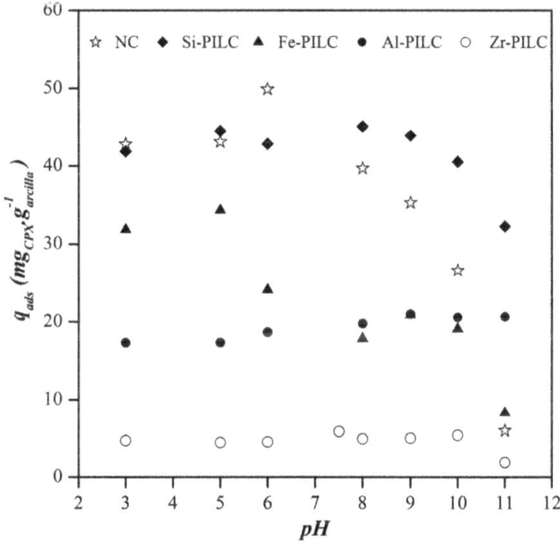

Figure 4. Effect of media pH on ciprofloxacin (CPX) adsorption.

The results for the pillared clays showed different behaviors according to the pillaring agent. The Fe-PILC exhibited the highest amount adsorbed at low pH, abruptly decreasing as the pH increases. This behavior can be explained by the interaction between the negative surface of the Fe-PILC and the positive species of CPX, favoring the adsorption at pH values lower than 5 [49]. Similar adsorption results were reported for rhodamine B and diclofenac on Fe-PILC [21,27]. For the Si-PILC, results showed no significant variations in the amount adsorbed at pH values below 8, and decreasing as the pH increases. However, the amounts adsorbed on the Si-PILC were much higher if compared to the adsorption of the other materials in the alkaline media. These results could suggest that Si-PILC has more available surface sites than the other PILC in this pH range, favoring the adsorption of CPX anionic species. On the other hand, the Si-PILC was the material with the highest micropores amount and narrow microporosity, both could be responsible for the increase in adsorption. These results are pretty interesting since there are no reports of Si-PILC being studied as adsorbents. The adsorption behaviors for Al- and Zr-PILC were similar, with no major differences in the amount adsorbed across the pH range. This behavior could suggest that for these types of materials, the adsorption mechanism is mainly governed by their porous structures limiting the access of the CPX molecule to the pillared structure. Analogous results were reported by Gil et al. [26] for the adsorption of orange II and methylene blue on Al- and Zr-PILC. That study took into account that the adsorption of a molecule occurs in pores with a diameter 1.3–1.8 times that of the solute. If this criterion is taken into account, and, since the adsorptive will diffuse into the porous structure of the adsorbent lengthwise, the minimum pore size for the adsorption of the CPX should be 1.31 nm. If that is the case, the adsorption results could be explained by Figure 3, where the aluminum and zirconium pillared clays are the materials with the lowest amount of pores that are higher than this size, similar to Fe-PILC, whereas the Si-PILC has higher amounts of pores in this range. This may indicate that the CPX molecule has more access to the pillars in the last material, which, in turn, could favor the interaction and adsorption between them.

The results obtained for the CPX adsorption on the PILC at different pH values suggest that some pillared materials would be optimal adsorbents of the CPX anionic specie. With that in mind and considering that the pH value of the natural water courses in the Alto Valle region is around 9, the adsorption and kinetics studies were carried out at pH of 10.

3.3. Adsorption Isotherms

The batch adsorption experiments were performed in the conditions previously mentioned, varying the initial concentrations between 18–500 mg L^{-1} and the contact time, now set for a period of 24 h. The adsorption studies were ran at pH 10, based on the results for CPX adsorption at different pH values and in order to evaluate the behavior of the pillared clays when the CPX anionic form is present. According to previous kinetic studies (not shown in this work), the optimal contact time was 24 h.

The adsorption isotherms obtained for CPX on different clays and their adjustments to the three models, are shown in Figure 5. Taking into account the Giles et al. classification [50], two different behaviors can be associated to the isotherm shapes illustrated there. The adsorption isotherms for CPX on pillared clays can be classified as high affinity type (H-type) and the one obtained for the natural clay was a Langmuir type (L-type) isotherm. In both of the cases, the isotherm shape is related to a progressive saturation of the solid surface due to the occupancy of the adsorbent surface sites, suggesting a high affinity of the adsorptive molecule toward the solid surface. The H-type isotherm is usually associated to the ionic solute adsorption where there is no strong competition between adsorptive and solvent molecules towards the surface of the solid [50,51]. This could be the result of a higher hydrophobicity being exhibited by the pillared clays in contrast with the natural clay. Another explanation is the presence of new adsorption sites in the pillared clays surface associated to the pillars. On the other hand, the L-type isotherm for the natural clay suggests a lower affinity of the anionic CPX species present toward the more negatively charged clay than the one that is observed for the PILC.

Freundlich, Langmuir, and Sips models were fitted to adsorption data obtained for all of the materials, and their fitting parameters are summarized in Table 2. The best fittings were obtained for the Sips model in all of the materials that were under study. This suggests heterogeneous systems that could result from the presence of different adsorption sites on the solid surface, the adsorbible species, or a combination of them. Similarly, when the parameter associated to the system heterogeneity in this model (n) is 1, the Sips equation becomes the Langmuir equation and the system can be considered to be a more homogeneous one. Thus, the n value nearest to 1 obtained for the natural clay suggests an adsorption system that is more homogeneous when compared to the values obtained for the PILC, which are greater than 2 in all of the cases. This indicates a more heterogeneous system, which is probably due to the PILC porous structure and the pillars presence. The highest adsorption capacity was obtained for Si-PILC and the lowest for Zr-PILC. The natural clay resulted in an intermediate adsorption capacity between two groups of pillared materials, Si- and Fe-PILC, which showed higher adsorption capacities and Al- and Zr-PILC, which were significantly lower. However, the affinity of the adsorbate for the solid surface is lower for the NC than it is for the pillared clays. Based on these results, the adsorption capacity for the NC could be explained by a hydrophobic effect of the solvent towards the organic molecule where the adsorption could be seen as a result of the repulsion of the organic molecule against the solvent from the solution [52]. The hydrophobic effect can be related to the k_{ow} value for the CPX, which is 1.9, indicating the hydrophobic character of the CPX molecule. After the molecule is in the solid-liquid interface, different adsorption short-range forces could be promoted between the CPX molecule and the solid surface, such as covalent and hydrophobic bonding, hydrogen bridges, steric, or orientation effects [52]. Additionally, the type L isotherm that is obtained is associated to a flatwise adsorption favoring van der Waals (π–π type) interactions between the aromatic fraction in the organic molecule and the siloxane surface of the clay material [48,50]. In the pillared clays, the hydrophobic effect could influence the adsorption the same way that it did for the NC. However, there are two additional factors that affect the adsorption capacity in these materials; the new adsorption sites that are generated by the pillars presence and the porous structure associated to them. The adsorption behaviors obtained are consistent with the results that are shown for the adsorption vs. pH, meaning that the highest adsorption capacity was obtained for the Si-PILC, whereas the lowest one was for Al- and Zr-PILC. The adsorption shown for the Si- and Fe-PILC could be due to a higher access of the CPX species to the porous structure when compared to the other PILC favoring its interaction with the different adsorption sites within the interlaminar region.

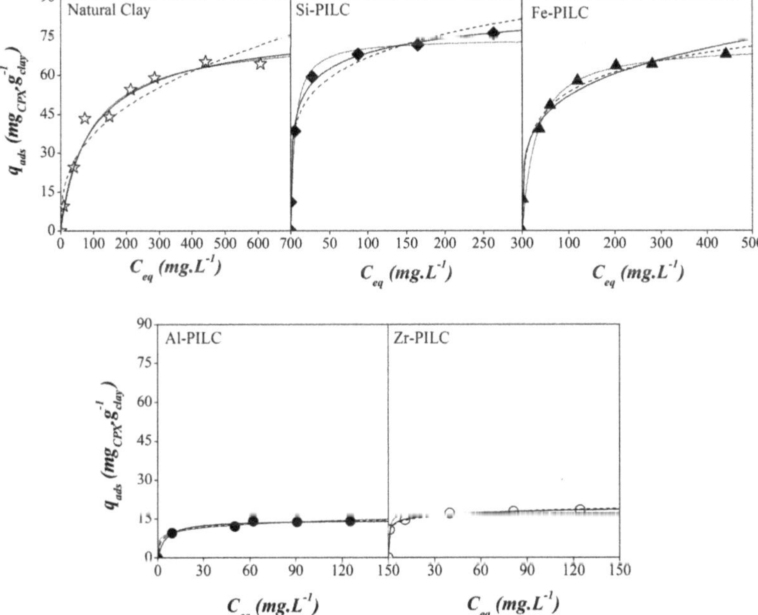

Figure 5. Experimental isotherms (symbols) and Langmuir (dash), Freundlich (dot) and Sips (straight) adjustments for the equilibrium adsorption data of CPX on natural clay (NC) and pillared clays (PILC).

The Scatchard plots obtained for all of the materials are shown in Figure 6. The R^2 values that are obtained for the whole range of data could suggest the presence of nonspecific or multi-type interactions between the adsorbate molecules and the surface sites. The R^2 values calculated were 0.906, 0.887, 0.856, 0.753 and 0.739 for AN, Al-PILC, Zr-PILC, Fe-PILC and Si-PILC, respectively. These values indicate a greater presence of nonspecific interactions for all PILC materials than there are for the NC, being the highest, the ones obtained for Fe- and Si-PILC. Furthermore, the Scatchard plots obtained for all the pillared clays can be considered as concave curves that are associated to a negative cooperative adsorption phenomenon, as well as surface heterogeneity [42,43]. As it can be seen in Figure 6, the Scatchard plots obtained for the PILC materials result in two independent sets of data, which individually arrange in a linear combination, where each one of them could be related to a type of affinity of the CPX specie to the surface. This may be the result of the presence of different adsorption sites in the clay surface causing the CPX^- to show high (H) and low (L) affinities towards the PILC surface [41,43]. These results could suggest that at an early stage of the adsorption process, the CPX^- interacts with the pillars either through the non-bonding electrons in its amine group or through the electrons of its carboxylate group, both with high affinity. However, this access is limited to a small amount of sites that become quickly saturated. Afterwards, the CPX species are adsorbed on the available sites of the clay surface by other types of low affinity interactions (i.e., van der Waals (π–π type) interactions, hydrogen bridges, etc.). On the other hand, the Scatchard plot that is obtained for the natural clay is a straight line, which is associated to an adsorption process where the solid surface only exhibits one type of site for the CPX anion to be adsorbed.

Table 2. Freundlich, Langmuir and Sips parameters for CPX adsorption on natural and pillared clay minerals.

Models	Units	NC	Si-PILC	Fe-PILC	Al-PILC	Zr-PILC
Freundlich model	k_F (mg g^{-1}(L mg^{-1})^n)	6.98	28.88	18.98	6.86	11.22
	n	2.75	5.49	4.57	6.45	9.48
	R^2	0.962	0.979	0.986	0.986	0.997
Langmuir model	q_m (mg g^{-1})	75.73	74.12	72.09	14.48	17.34
	K (L mg^{-1})	0.01	0.19	0.03	0.20	1.78
	R^2	0.993	0.971	0.966	0.986	0.975
Sips model	q_m (mg g^{-1})	80.82	100.60	122.10	17.78	25.20
	b (L mg^{-1})	0.01	0.07	0.01	0.14	0.36
	n	1.13	2.56	2.76	2.07	3.85
	R^2	0.993	0.996	0.991	0.987	0.999

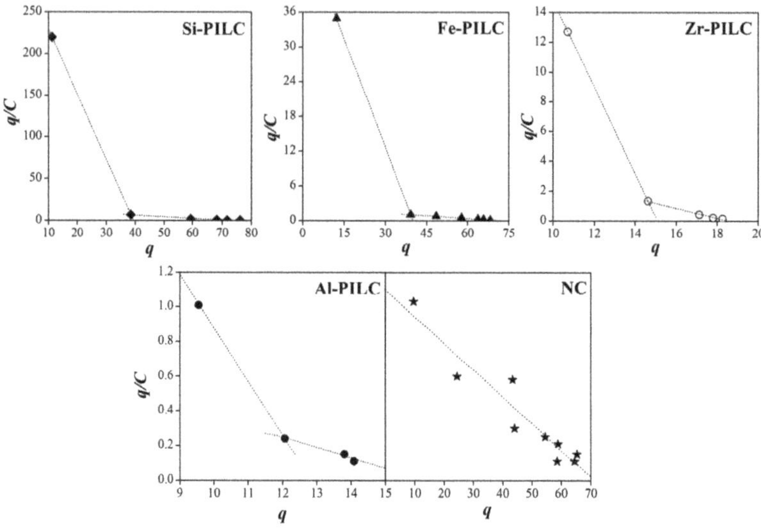

Figure 6. Scatchard plots derived for adsorption data obtained at pH 10 for the five materials.

The results show a close relationship between the adsorption capacity of the PILC materials and their porous structure, as well as the influence of the micropores size on the CPX^- access to the interlaminar space for its subsequent adsorption. However, in order to obtain complementary information about the possible contribution of the mesopores present in the materials structure on the CPX^- adsorption, textural properties were evaluated. Table 3 shows the values of the cumulative volumes for the pillared clays when considering three ranges; (1) the amount of micropores whose size is higher than 2 nm ($V_{\mu p} < 2$ nm), (2) the mesopores ranging between 2–10 nm, and the (3) the mesopores between 10 and 50 nm. The values show that the materials with the highest adsorption capacities (Si- and Fe-PILC) also have the highest amount of mesopores with a size of lower than 10 nm. This indicates that the mesopores are playing an important role in the CPX adsorption on pillared clays under the studied conditions. This type of pores might be more accessible for the molecule and represent adsorption sites for the kind of interactions mentioned earlier.

Table 3. Cumulative volumes for the pillared clays.

Pillared Clays	$V_{\mu p}$ (<2 nm)	V_{mp} (2–10 nm)	V_{mp} (10–50 nm)	V_T
Si-PILC	0.09	0.18	0.04	0.31
Fe-PILC	0.03	0.08	0.06	0.17
Al-PILC	0.10	0.04	0.04	0.18
Zr-PILC	0.06	0.05	0.06	0.17

Table 4 summarizes the results reported for other authors for the adsorption of CPX on different adsorbents, and they are compared to the results obtained in this work. It shows that most of the reported studies were carried out at pH values lower than 7 and under these conditions the materials with highest adsorption capacity of CPX are the clay minerals. These results probably are because at those pH values the cationic species of CPX is present and it has high affinity for the negative clay minerals surfaces. However, there are less reports of the CPX adsorption at pH values higher than 7 when the anionic species is present. In this sense, the adsorption capacity of the natural clay mineral was lower than the results showed for the pillared clays, suggesting that these materials could be good adsorbents for anionic species.

Table 4. CPX adsorption capacities for different materials.

Adsorbent	$q_{m,CPX}$ (pH) (mg g^{-1})	Reference
Aluminum hydrous oxide	14.72 (7)	[14]
Iron hydrous oxide	25.76 (7)	
Ca^{2+}-montmorillonite (Saz)	330 (4–5.5)	[19]
Activated carbon	231 ($\approx$7)	
Carbon nanotubes	135 ($\approx$7)	[8]
Carbon xerogel	112 ($\approx$7)	
kaolinite	6.99 (5–6)	[53]
Illite	33 (4–5.5)	
Rectorie	135 (4–5.5)	[18]
Bentonite	147 (4.5)	[15]
Birnessite	80.96 (5–6)	[16]
Montmorillonite	332.8 (3)	[17]
	138.7 (6)	
	71.6 (7.5)	
	80.82 (10)	
Graphene Oxide	379 (5)	[11]
CMK-3	281.47 (<7)	
CMK-3 modified	369.34 (<7)	[10]
Bamboo-based carbon	153.17 (<7)	
Bamboo-based carbon modified	237.44 (<7)	
Multi-walled nanotubes	194 (4)	[12]
Si-PILC	100.6 (10)	This work
Fe-PILC	122.1 (10)	This work
Al-PILC	17.78 (10)	This work
Zr-PILC	25.20 (10)	This work

3.4. Evidences of CPX Interactions with PILC

Looking for evidences of the interactions between the CPX species and the pillared clays surface, FTIR (Fourier-transform infrared spectroscopy) spectra of adsorbed CPX on the PILC (adsorption complex) were obtained. The adsorption complexes were studied for the saturated points in the adsorption isotherms, after the centrifugation step, each one of them was dried at room temperature and the FTIR of the resultant solids were obtained. The spectra were compared with those obtained for the CPX (pure) and for the PILC materials.

Figure 7 shows the resultant spectra for all of the samples CPX, PILC materials, and adsorption complexes. The vibration bands that were obtained for the pure CPX are comparable with the ones that were obtained in the previous work [17]. In the same sense, the vibration bands that are associated to the interaction between the CPX species and different solids and metals have been previously reported [14,54]. Taking these works into account, the bands at 1264 and 1700 cm^{-1} found in the CPX spectrum are assigned to the protonation of the carboxylic group and the stretching of its carbonyl group, respectively. After the adsorption, the first band shifted to 1274 cm^{-1} for all of the adsorption complexes, suggesting that the carboxylate group is involved in the adsorption process. This possibly occurs by Lewis acid-base interactions between its electrons and the metallic atoms that are available in the solid surface. The second band is missing, whereas two bands appear at 1636 and 1490 cm^{-1} for the adsorption complexes that are assigned to the asymmetric and the symmetric stretch of the coordinated carboxylate group of the CPX molecule. The presence of these last bands could be associated to the CPX$^-$ acting as a mononuclear bidentate ligand in the adsorption complexes where the oxygen of the carbonyl group, belonging to the quinolone moiety, and one of the oxygen atoms of the carboxylate group interact with the available metallic atoms on the pillared clays surface. Those kinds of complexes have been proposed previously by other authors for the adsorption of CPX on aluminum and iron hydrous oxides [14]. In view of this evidence, and considering the previous works, the possible structure for the interaction between CPX and the metal atoms on the solid surface is represented in Figure 8.

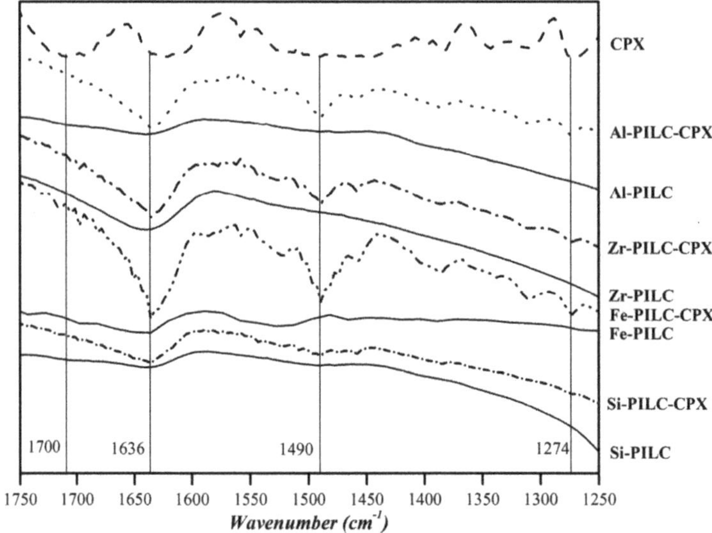

Figure 7. FTIR (Fourier Transform infrared spectroscopy) spectra of CPX, PILC, and the adsorption complexes obtained.

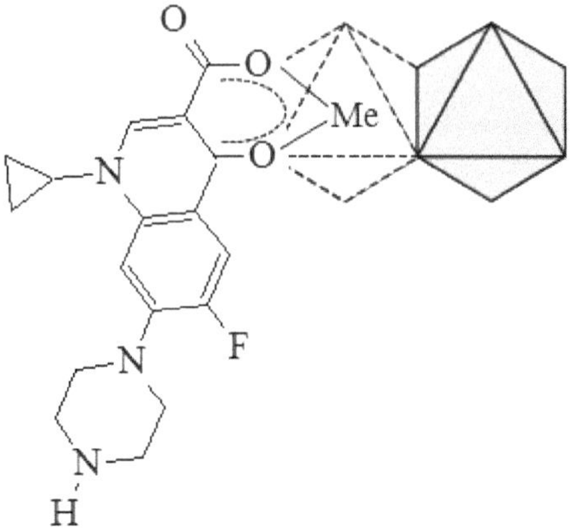

Figure 8. Representation of the structure proposed for the interaction between CPX^- and atoms of metals on the PILC surface.

4. Conclusions

In this study, different pillared clay minerals were evaluated as adsorbents of ciprofloxacin in basic conditions. The highest CPX adsorption capacity was obtained for the Si- and Fe-PILC at the studied conditions and it could be related to both, the presence of micro and mesoporous with sizes greater than the Al- and Zr-PILC, and the new adsorption sites that are generated for the metal atom in the pillars. Adsorption data evidenced that there is a strong relationship between the porous structure of the pillared clay and their adsorption capacity, suggesting that the latter could be mainly related to the access of the CPX molecule to the pillars within the PILC interlaminar region. Additionally, the results suggest that the adsorption mechanism for the CPX on the PILC involves a first moment governed for the hydrophobic effect on CPX^- followed by the adsorption that is caused by the inner-sphere complexes formation, as well as by the van der Waals interactions between CPX^- and the PILC sites surface.

Acknowledgments: The authors gratefully acknowledge the Universidad Nacional del Comahue, Universidad Nacional San Luis, CONICET and ANPCyT (Agencia Nacional de Promoción Científica y Tecnológica) for the financial support.

Author Contributions: Maria Eugenia Roca Jalil designed, performed the experiments, analyzed data and wrote the paper. Miria Baschini and Karim Sapag cooperated with the data analysis and manuscript elaboration. All the authors read and approved the final manuscript.

Conflicts of Interest: The authors declare no conflict of interest.

References

1. Dietrich, D.R.; Hitzfeld, B.C.; O'Brien, E. Toxicology and Risk assessment of pharmaceuticals. In *Organic Pollutants in the Water Cycle*; Reemtsma, T., Jekel, M., Eds.; John Wiley & Sons: Weinheim, Germany, 2006; pp. 287–309; ISBN 9783527608775.
2. Halling-Sørensen, D.; Holten Lützhøft, H.C.; Andersen, H.R.; Ingerslev, F. Environmental risk assessment of antibiotics: Comparison of mecillinam, trimethoprim and ciprofloxacin. *J. Antimicrob. Chemother.* **2000**, *46*, 53–58. [CrossRef] [PubMed]
3. Kümmerer, K. The presence of pharmaceuticals in the environment due to human use-present knowledge and future challenges. *J. Environ. Manag.* **2009**, *90*, 2354–2366. [CrossRef] [PubMed]
4. Ahmed, M.B.; Zhou, J.L.; Ngo, H.H.; Guo, W. Adsorptive removal of antibiotics from water and wastewater: Progress and challenges. *Sci. Total Environ.* **2015**, *532*, 112–126. [CrossRef] [PubMed]

5. Akhtar, J.; Amin, N.A.S.; Shahzad, K. A review on removal of pharmaceuticals from water by adsorption. *Desalin. Water Treat.* **2016**, *57*, 12842–12860. [CrossRef]
6. Grassi, M.; Kaykioglu, G.; Belgiorno, V.; Lofrano, G. Removal of Emerging contaminants from water and wastewater by adsorption process. In *Emerging Compounds Removal from Wastewater. Natural and Solar Based Treatments*; Lofrano, G., Ed.; Springer: Dordrecht, The Netherlands, 2012; pp. 15–37.
7. Balarak, D.; Mostafapour, F.K.; Bazrafshan, E.; Saleh, T.A. Studies on the adsorption of amoxicillin on multi-wall carbon nanotubes. *Water Sci. Technol.* **2017**, *75*, 1599–1606. [CrossRef] [PubMed]
8. Carabineiro, S.A.C.; Thavorn-Amornsri, T.; Pereira, M.F.R.; Figueiredo, J.L. Comparison between activated carbon, carbon xerogel and carbon nanotubes for the adsorption of the antibiotic ciprofloxacin. *Catal. Today* **2012**, *186*, 29–34. [CrossRef]
9. Calisto, V.; Ferreira, C.I.; Oliveira, J.A.; Otero, M.; Esteves, V.I. Adsorptive removal of pharmaceuticals from water by commercial and waste-based carbons. *J. Environ. Manag.* **2015**, *152*, 83–90. [CrossRef] [PubMed]
10. Peng, X.; Hu, F.; Lam, F.L.; Wang, Y.; Liu, Z.; Dai, H. Adsorption behavior and mechanisms of ciprofloxacin from aqueous solution by ordered mesoporous carbon and bamboo-based carbon. *J. Colloid Interface Sci.* **2015**, *460*, 349–360. [CrossRef] [PubMed]
11. Chen, H.; Gao, B.; Li, H. Removal of sulfamethoxazole and ciprofloxacin from aqueous solutions by graphene oxide. *J. Hazard. Mater.* **2015**, *282*, 201–207. [CrossRef] [PubMed]
12. Yu, F.; Sun, S.; Han, S.; Zheng, J.; Ma, J. Adsorption removal of ciprofloxacin by multi-walled carbon nanotubes with different oxygen contents from aqueous solutions. *Chem. Eng. J.* **2016**, *285*, 588–595. [CrossRef]
13. Liang, Z.; Zhaob, Z.; Sun, T.; Shi, W.; Cui, F. Adsorption of quinolone antibiotics in spherical mesoporous silica: Effects of the retained template and its alkyl chain length. *J. Hazard. Mater.* **2016**, *305*, 8–14. [CrossRef] [PubMed]
14. Gu, C.; Karthileyan, G. Sorption of the antimicrobial ciprofloxacin to aluminum and iron hydrous oxides. *Environ. Sci. Technol.* **2005**, *39*, 9166–9173. [CrossRef] [PubMed]
15. Genç, N.; Dogan, E.C.; Yurtsever, M. Bentonite for ciprofloxacin removal from aqueous solution. *Water Sci. Technol.* **2013**, *68*, 848–855. [CrossRef] [PubMed]
16. Jiang, W.-T.; Chang, P.-H.; Wang, Y.-S.; Tsai, Y.; Jean, J.-S.; Li, Z.; Krukowski, K. Removal of ciprofloxacin from water by birnessite. *J. Hazard. Mater.* **2013**, *250–251*, 362–369. [CrossRef] [PubMed]
17. Roca Jalil, M.E.; Baschini, M.; Sapag, K. Influence of pH and antibiotic solubility on the removal of ciprofloxacin from aqueous media using montmorillonite. *Appl. Clay Sci.* **2015**, *114*, 69–76. [CrossRef]
18. Wang, C.-J.; Li, Z.; Jiang, W.-T. Adsorption of ciprofloxacin on 2:1 dioctahedral clay minerals. *Appl. Clay Sci.* **2011**, *53*, 723–728. [CrossRef]
19. Wang, C.-J.; Li, Z.; Jiang, W.-T.; Jean, J.-S.; Liu, C.-C. Cation Exchange interaction between antibiotic ciprofloxacin and montmorillonite. *J. Hazard. Mater.* **2010**, *183*, 309–314. [CrossRef] [PubMed]
20. Putra, E.K.; Pranowo, R.; Sunarso, J.; Indraswati, N.; Ismadji, S. Performance of activated carbon and bentonite for adsorption of amoxicillin from wastewater: Mechanisms, isotherms and kinetics. *Water Res.* **2009**, *43*, 2419–2430. [CrossRef] [PubMed]
21. Mabrouki, H.; Akretche, D.E. Diclofenac potassium removal from water by adsorption on natural and Pillared Clay. *Desalin. Water Treat.* **2016**, *57*, 6033–6043. [CrossRef]
22. Parolo, M.E.; Savini, M.; Valles, J.; Baschini, M.; Avena, M. Tetracycline adsorption on montmorillonite: Effects of pH and ionic strength. *Appl. Clay Sci.* **2008**, *40*, 179–186. [CrossRef]
23. Wu, H.; Xie, H.; He, G.; Guan, Y.; Zhang, Y. Effects of the pH and anions on the adsorption of tetracycline on iron-montmorillonite. *Appl. Clay Sci.* **2016**, *119*, 161–169. [CrossRef]
24. Al-Khalisy, R.S.; Al-Haidary, A.M.A.; Al-Dujaili, A.H. Aqueous phase adsorption of cephalexin onto bentonite and activated carbon. *Sep. Sci. Technol.* **2010**, *45*, 1286–1294. [CrossRef]
25. Gil, A.; Korili, S.A.; Vicente, M.A. Recent Advances in the control and characterization of the porous structure of pillared clay catalysts. *Catal. Rev.* **2008**, *50*, 153–221. [CrossRef]
26. Gil, A.; Assis, F.C.C.; Albeniz, S.; Korili, S.A. Removal of dyes from wastewaters by adsorption on Pillared clays. *Chem. Eng. J.* **2011**, *168*, 1032–1040. [CrossRef]
27. Hou, M.-F.; Ma, C.-X.; Zhang, W.-D.; Tang, X.-Y.; Fan, Y.-N.; Wan, H.-F. Removal of rhodamine B using iron-pillared bentonite. *J. Hazard. Mater.* **2011**, *186*, 1118–1123. [CrossRef] [PubMed]

28. Liu, Y.N.; Dong, C.; Wei, H.; Yuan, W.; Li, K. Adsorption of levofloxacin onto an iron-pillared montmorillonite (clay mineral): Kinetics, equilibrium and mechanism. *Appl. Clay Sci.* **2015**, *118*, 301–307. [CrossRef]

29. Manohar, D.M.; Noelite, B.F.; Anirudhan, T.S. Removal of Vanadium (IV) from aqueous solutions by adsorption process with aluminum-pillared bentonite. *Ind. Eng. Chem. Res.* **2005**, *44*, 6676–6684. [CrossRef]

30. Mishra, T.; Mahato, D.K. A comparative study on enhanced arsenic (V) and arsenic (III) removal by iron oxide and manganese oxide pillared clays from ground water. *J. Environ. Chem. Eng.* **2016**, *4*, 1224–1230. [CrossRef]

31. Molu, Z.B.; Yurdakoç, K. Preparation and characterization of aluminum Pillared K10 and KSF for adsorption of thimethoprim. *Microporous Mesoporous Mater.* **2010**, *127*, 50–60. [CrossRef]

32. Roca Jalil, M.E.; Baschini, M.; Rodríguez-Castellón, E.; Infantes-Molina, A.; Sapag, K. Effect of the Al/clay ratio on the thiabendazole removal by aluminum pillared clays. *Appl. Clay Sci.* **2014**, *87*, 245–263. [CrossRef]

33. Roca Jalil, M.E.; Vieria, R.S.; Azevedo, D.; Baschini, M.; Sapag, M. Improvement in the adsorption of thiabendazole by using aluminum pillared clays. *Appl. Clay Sci.* **2013**, *71*, 55–63. [CrossRef]

34. Yan, L.G.; Xu, Y.Y.; Yu, H.Q.; Xin, X.D.; Wei, Q.; Du, B. Adsorption of phosphate from aqueous solution by hydroxy-aluminum, hydroxy-iron and hydroxy-iron–aluminum pillared bentonites. *J. Hazard. Mat.* **2010**, *179*, 244–250. [CrossRef] [PubMed]

35. Han, Y.S.; Matsumoto, H.; Yamanaka, S. Preparation of new silica sol-based Pillared clays with high Surface area and high thermal stability. *Chem. Mater.* **1997**, *9*, 2013–2018. [CrossRef]

36. Yamanaka, S.; Doi, T.; Sako, S.; Hattori, M. High surface area solids obtained by intercalation of iron oxide pillars in montmorillonite. *Mater. Res. Bull.* **1984**, *19*, 161–168. [CrossRef]

37. Farfan-Torres, E.M.; Dedeycker, O.; Grange, P. Zirconium pillared clays. Influence of basic polymerization of the precursor on their structure and stability. *Stud. Surf. Sci. Catal.* **1991**, *63*, 337–343.

38. Rouquerol, F.; Rouquerol, J.; Sing, K.; Llewellyn, P.; Maurin, G. Adsorption by powders and porous solids. In *Principles Methodology and Applications*, 2nd ed.; Elsevier Academic Press: Amsterdam, The Netherlands, 2013; ISBN 978-0-08-097035-6.

39. Villarroel-Rocha, J.; Barrera, D.; García Blanco, A.A.; Roca Jalil, M.E.; Sapag, K. Importance of the α-plot Method in the characterization of nanoporous materials. *Adsorpt. Sci. Technol.* **2013**, *31*, 165–183. [CrossRef]

40. Febrianto, J.; Kosasih, A.N.; Sunarso, J.; Ju, Y.; Indraswati, N.; Ismadji, S. Equilibrium and kinetic studies in adsorption of heavy metals using biosorbent: A summary of recent studies. *J. Hazard. Mater.* **2009**, *162*, 616–645. [CrossRef] [PubMed]

41. Anirudhan, T.S.; Suchithra, P.S. Equilibrium, kinetic and thermodynamic modeling for the adsorption of heavy metals onto chemically modified hydrotalcite. *Indian J. Chem. Technol.* **2010**, *17*, 247–259.

42. Dahlquist, F.W. *The Meaning of Scatchard and Hill Plots in: Methods of Enzymology*; Hirs, C.H.W., Timasheff, S.N., Eds.; Academic Press: New York, NY, USA, 1978; Volume 48, pp. 270–299.

43. Gerente, C.; Couespel du Mesnil, P.; Andrès, Y.; Thibault, J.-F.; Le Cloirec, P. Removal of metal ions from aqueous solution on low cost natural polysaccharides Sorption mechanism approach. *React. Funct. Polym.* **2000**, *46*, 135–144. [CrossRef]

44. Gezici, O.; Kara, H.; Ayar, A.; Topkafa, M. Sorption behavior of Cu(II) ions on insolubilized humic acid under acidic conditions: An application of Scatchard plot analysis in evaluating the pH dependence of specific and nonspecific bindings. *Sep. Purif. Technol.* **2007**, *55*, 132–139. [CrossRef]

45. Komadel, P.; Madejová, J. Acid activation of Clay Minerals. In *Handbook of Clay Science Developments in Clay Science, Part A: Fundamentals*, 2nd ed.; Bergaya, F.G., Lagaly, G., Eds.; Elsevier Ltd. Press: Amsterdam, The Netherlands, 2013; Volume 2, pp. 385–408; ISBN 9780080993713.

46. Mendioroz, S.; Pajares, A.; Benito, I.; Pesquera, C.; González, F.; Blanco, C. Texture evolution of montmotillonite under progressive acid treatment: Change from H3 to H2 type of hysteresis. *Langmuir* **1987**, *3*, 676–681. [CrossRef]

47. Thommes, M.; Kaneko, K.; Neimark, A.V.; Olivier, J.P.; Rodriguez-Reinoso, F.; Rouquerol, J.; Sing, K.S. Physisorption of gases, with special reference to the evaluation of surface area and pore size distribution (IUPAC Technical Report). *Pure Appl. Chem.* **2015**, *87*, 1051–1069. [CrossRef]

48. Lagaly, G.; Ogawa, M.; Dékány, I. Clay mineral organic interactions. In *Handbook of Clay Science-Developments in Clay Science*; Bergaya, F., Theng, B.K.G., Lagaly, G., Eds.; Elsevier Press: Amsterdam, The Netherlands, 2013; Volume 5A, pp. 435–505; ISBN 978-0-08-044183-2.

49. Marco-Brown, J.L.; Barbosa-Lema, C.M.; Torres Sanchez, R.M.; Mercader, R.C.; dos Santos Afonso, M. Adsorption of picloram herbicide on iron oxide pillared montmorillonite. *Appl. Clay Sci.* **2012**, *58*, 25–33. [CrossRef]

50. Giles, C.H.; Smith, D.; Huitson, A. A general treatment and classification of the solute adsorption isotherm. I. Theoretical. *J. Colloid Interface Sci.* **1974**, *47*, 755–765. [CrossRef]

51. Limousin, G.; Gaudet, J.P.; Charlet, L.; Szenknect, S.; Barthes, V.; Krimissa, M. Sorption isotherms: A review on physical bases, modeling and measurement. *Appl. Geochem.* **2007**, *22*, 249–275. [CrossRef]

52. Stumm, W. *Chemistry of the Solid-Water Interface. Processes at the Mineral-water and Particle-Water Interface in Natural Systems*; John Wiley & Son Inc. Press: New York, NY, USA, 1992; ISBN 0471576727.

53. Li, Z.; Hong, H.; Liao, L.; Ackley, C.J.; Schulz, L.A.; MacDonald, R.A.; Emard, S.M. A mechanistic study of ciprofloxacin removal by kaolinite. *Colloid Surface B* **2011**, *88*, 339–344. [CrossRef] [PubMed]

54. Al-Mustafa, J.; Tashtoush, B. Iron (II) and iron (III) perchlorate complexes of ciprofloxacin and norfloxacin. *J. Coord. Chem.* **2003**, *56*, 113–124. [CrossRef]

Article

Novel Montmorillonite/TiO$_2$/MnAl-Mixed Oxide Composites Prepared from Inverse Microemulsions as Combustion Catalysts

Bogna D. Napruszewska [1], Alicja Michalik-Zym [1], Melania Rogowska [1], Elżbieta Bielańska [1], Wojciech Rojek [1], Adam Gaweł [2], Monika Wójcik-Bania [2], Krzysztof Bahranowski [2] and Ewa M. Serwicka [1,*]

[1] Jerzy Haber Institute of Catalysis and Surface Chemistry, Niezapominajek 8, 30-239 Krakow, Poland; ncnaprus@cyf-kr.edu.pl (B.D.N.); ncmichal@cyf-kr.edu.pl (A.M.-Z.); melania.rogowska@gmail.com (M.R.); ncbielan@cyf-kr.edu.pl (E.B.); ncrojek@cyf-kr.edu.pl (W.R.)

[2] Faculty of Geology, Geophysics and Environmental Protection, AGH University of Science and Technology, al. Mickiewicza 30, 30-059 Krakow, Poland; agawel@agh.edu.pl (A.G.); wojcikm@agh.edu.pl (M.W.-B.); bahr@agh.edu.pl (K.B.)

* Correspondence: ncserwic@cyf-kr.edu.pl; Tel.: +48-12-6395-118

Received: 30 October 2017; Accepted: 17 November 2017; Published: 19 November 2017

Abstract: A novel design of combustion catalysts is proposed, in which clay/TiO$_2$/MnAl-mixed oxide composites are formed by intermixing exfoliated organo-montmorillonite with oxide precursors (hydrotalcite-like in the case of Mn-Al oxide) obtained by an inverse microemulsion method. In order to assess the catalysts' thermal stability, two calcination temperatures were employed: 450 and 600 °C. The composites were characterized with XRF (X-ray fluorescence), XRD (X-ray diffraction), HR SEM (high resolution scanning electron microscopy, N$_2$ adsorption/desorption at −196 °C, and H$_2$ TPR (temperature programmed reduction). Profound differences in structural, textural and redox properties of the materials were observed, depending on the presence of the TiO$_2$ component, the type of neutralization agent used in the titania nanoparticles preparation (NaOH or NH$_3$ (aq)), and the temperature of calcination. Catalytic tests of toluene combustion revealed that the clay/TiO$_2$/MnAl-mixed oxide composites prepared with the use of ammonia showed excellent activity, the composites obtained from MnAl hydrotalcite nanoparticles trapped between the organoclay layers were less active, but displayed spectacular thermal stability, while the clay/TiO$_2$/MnAl-mixed oxide materials obtained with the aid of NaOH were least active. The observed patterns of catalytic activity bear a direct relation to the materials' composition and their structural, textural, and redox properties.

Keywords: montmorillonite/hydrotalcite composite; montmorillonite/titania composite; organoclay; inverse micelle; Mn-Al mixed oxide; combustion catalysts

1. Introduction

Widespread emission of volatile organic compounds (VOCs) to the atmosphere, occurring chiefly as a result of industrial and transportation activities, is regarded as a major global environmental hazard, due to the toxic, mutagenic, and carcinogenic nature of the pollutants, and to their role in the formation of the photochemical smog. Among the different techniques for abatement of VOCs, neutralization based on catalytic combustion is considered particularly appropriate, owing to its low operational costs and high destruction efficiency [1]. Catalyst design is usually based on noble metals or on transition metal oxides, the latter being attractive due to their lower price. It has been repeatedly demonstrated that efficient VOCs combustion catalysts can be developed from clay minerals of anionic and/or cationic character (e.g., [2–19]). Recently, we proposed a novel strategy for preparing VOCs combustion catalysts, in which the catalytically active Mn-Al oxide nanoparticles, obtained from hydrotalcite (Ht) precursor synthesized by an inverse microemulsion method, are trapped between randomly oriented layers of smectite (Laponite) [19]. In such a composite the active phase forms uniform grains of nanometer dimensions, well separated by clay layers, which hinders sintering phenomena. Moreover, termination of clay basal faces with the chemically inert silica sheet prevents catalyst deactivation via reaction with the support.

In the present work we propose further modification of the catalyst design by incorporation of Mn-Al inverse micelles into the smectite matrix loaded previously with titania. TiO_2 belongs to the most frequently studied catalyst supports, and in many cases it has been shown that its use is beneficial for catalytic activity in total oxidation reactions [20–24]. However, rather than using conventional microporous Ti-pillared clay (Ti-PILC), we employed novel materials, recently developed in our laboratory, referred to as Ti-IMEC composites, formed from exfoliated organo-montmorillonite and inverse microemulsion containing Ti oxo-hydroxy species [25]. The composites contain larger TiO_2 particles and are therefore mesoporous, which makes them texturally more suitable for entrapment of inverse micelles with Mn-Al Ht-like precursor. In the present study montmorillonite was chosen as the clay component. Toluene was used as a model VOC compound.

2. Materials and Methods

2.1. Materials

Active phase precursor in the form of MnAl Ht-like nanoparticles was prepared following the double-microemulsion method proposed by Bellezza et al. [26] and described in detail previously [19]. Inverse microemulsion consists of very small droplets of water dispersed in a continuous oil phase. The water droplets are stabilized by the presence of surfactant and co-surfactant. When two microemulsions, each containing reactants required for Ht formation, are mixed, the intermicellar exchange of solutes leads to product precipitation within the limited space of the aqueous micellar core, yielding nano-size Ht particles. Briefly, two inverse microemulsions of Ht-forming reagents were prepared, one from the aqueous solution of $Mn(NO_3)_2$ and $Al(NO_3)_3$, with the intended Mn:Al = 3, the other from NH_3 (aq). Each aqueous solution was dispersed in the organic medium based on isooctane as oil phase, cetyltrimethylammonium bromide (CTABr) as surfactant, and n-butanol as co-surfactant, and stirred till the mixtures turned into stable transparent liquids. Equal volumes of the two microemulsions were mixed at 70 °C to enable precipitation of the Ht-like phase, referred to as MnAl(im). The suspensions of MnAl(im) precursor in the organic mother liquor were further used for the preparation of composites. For the sake of comparison, the Ht-like material of similar stoichiometry was prepared by the standard co-precipitation method and is denoted MnAl(st).

The starting montmorillonite clay used for the preparation of composites was the sodium form of the less than 2 μm particle size fraction separated by sedimentation from Jelšový Potok bentonite (Slovakia), denoted Na-mt. The material was transformed into an organic derivative, referred to as CTA-Mt, by a routine cation exchange procedure with the CTABr aqueous solution.

The TiO_2-containing catalysts were obtained via Ti-IMEC intermediates prepared following the procedure described in detail previously [25]. Briefly, the CTA-Mt form was subjected to an exfoliation treatment [27] by dispersing in 1-hexanol and ultrasonication. Inverse Ti-containing microemulsion was prepared from an aqueous solution of Ti species obtained by controlled hydrolysis of $TiCl_4$ [28] added to 1-hexanol as oil phase and CTABr as surfactant. The system was stirred till the mixture turned into a stable transparent liquid. Thus formed Ti-containing microemulsion was added drop-wise to CTA-Mt dispersion in 1-hexanol, at 60 °C, in the amount providing a Ti loading of ca. 40 wt%. Subsequently, the pH was raised to 7 either by drop-wise addition of 1 M NaOH as described in reference [25] and the dispersion stirred for further 0.5 h. Alternatively, to avoid introduction of sodium into the composite material, raising of the pH was carried out using 25%

NH$_3$ (aq) solution. The product obtained by neutralization with NaOH is further referred to as Ti-IMEC$_{NaOH}$, the one neutralized with ammonia, as Ti-IMEC$_{NH3}$. The final composites were obtained by mixing Ti-IMEC$_{NaOH}$ or Ti-IMEC$_{NH3}$ with MnAl(im), all components used as suspensions in their organic mother liquors. For each Ti-IMEC preparative route (NaOH or NH$_3$) composites with two different MnAl(im) contents were obtained, the intended loadings corresponding approximately to 1:9 and 1:3 MnAl(im)/Ti-IMEC weight ratios. The mixtures were vigorously stirred for 18 h, followed by 1 h ultrasonication. Finally, the precipitates were washed, initially with a 1:1 mixture of chloroform and ethanol, followed by 1:1 mixture of ethanol and water, ending with distilled water. The products were dried by lyophilization and calcined for 4 h at 450 °C or 600 °C. The catalysts were denoted MnAl(im)/Ti-IMEC$_{NaOH}$-x-temp, or MnAl(im)/Ti-IMEC$_{NH3}$-x-temp, where "x" was I or II, depending on the load of MnAl(im) component, and "temp" was the temperature of calcination, i.e., the sample signature MnAl(im)/Ti-IMEC$_{NaOH}$-II-450 means that the catalyst was obtained from Ti-IMEC neutralized with NaOH, containing larger loading of MnAl(im) component and was calcined at 450 °C.

In order to check on the role of titania component, a composite without TiO$_2$ addition was obtained, according to the procedure described in reference [19], by combining suspensions of organoclay with MnAl(im). Briefly, the CTA-Mt was dispersed in isopropanol and added to the as received suspension of MnAl(im) in organic mother liquor, in relative amounts corresponding to the higher of the two active phase loadings used in the preparation of TiO$_2$-containing composites. After vigorous stirring at 20 °C the composite was filtered off, lyophilized and calcined for 4 h at 450 °C or 600 °C. The resulting catalysts are referred to as MnAl(im)/CTA-Mt-450 and MnAl(im)/CTA-Mt-600. Finally, for the sake of assessing the purposefulness of the catalyst synthesis from inverse microemulsion/organoclay mixture, yet another composite was prepared, of the clay/active phase proportion similar to that of MnAl(im)/CTA-Mt, this time via a simple totally inorganic route, by mixing appropriate amounts of aqueous suspension of MnAl hydrotalcite, obtained by the standard co-precipitation method, and referred to as MnAl(st), and the aqueous dispersion of the sodium form of montmorillonite, Na-Mt. Both components were stirred vigorously, filtered off, lyophilized and calcined at 450 and 600 °C for 4 h. The samples are referred to as MnAl(st)/Na-Mt-450 and MnAl(st)/Na-Mt-600.

2.2. Methods

Powder X-ray diffraction (XRD) patterns were recorded using X'Pert PRO MPD diffractometer (PANalytical, Holland) with CuKα radiation. The crystallite sizes of anatase modification of TiO$_2$ were estimated as an average of Scherrer calculations carried out for the (101) and (200) peaks, and of Mn$_3$O$_4$ for the (112), (103), and (211) reflections.

Chemical composition of the investigated solids was determined with a ZSX Primus II spectrometer (Rigaku, Japan) with a Rh anode as X-ray source, using a calibration based on certified reference materials.

High magnification SEM images were recorded for the uncoated samples deposited on 200 Mesh copper grids covered with carbon support film, using a JSM-7500F Field Emission Scanning Electron Microscope (SEM) (JEOL, Japan).

Temperature programmed reduction (TPR) was performed in a quartz U-shaped tubular reactor (home-made). About 0.015 g of sample was used. The reducing gas was a mixture of 5 vol% H$_2$ in Ar (Linde, H$_2$ 5% in Ar), at a total flow rate of 30 mL·min^{-1}. The temperature was increased at a rate of 10 °C·min^{-1} from room temperature to 700 °C. The TPR profiles were recorded using a thermal conductivity detector (TCD).

Textural parameters were derived from N$_2$ adsorption/desorption measurements performed at −196 °C with the use of an AUTOSORB 1 instrument (Quantachrome Instruments, USA). Prior to measurement, the samples were outgassed at 200 °C for 3 h. Specific surface areas were calculated according to the Brunauer–Emmett–Teller method (S$_{BET}$) in the relative pressure range 0.02–0.04. The total pore volume (V$_{tot}$) was calculated from the amount of N$_2$ adsorbed at a relative vapor pressure

$p/p_0 = 0.996$. The mean diameters of all pores (D^{av}) were evaluated using the Gurvitch formula $D^{av} = 4V_{tot}/S_{BET}$. Pore size distribution (PSD) profiles were determined by the NL DFT method.

Catalytic combustion of toluene was carried out in a fixed-bed flow quartz reactor of 10 mm inner diameter, loaded with ca. 0.5 g of a catalyst (particle size 0.3–0.5 mm), in the temperature range 100–400 °C. Toluene at 500 ppm concentration was fed to the flow by passing air through a saturator, at GHSV of 10,000 h^{-1}. The only reaction products were CO_2 and water. Toluene consumption was measured by GC-FID (SRI 8610A) and CO_2 evolution by GC-FID (SRI 310).

3. Results and Discussion

3.1. Characterization of Composites

The chemical composition of the synthesized materials is shown in Table 1. The data indicate that in the titania-containing composites the clay component constitutes about half of the total catalyst weight, with a Ti/Si atomic ratio of ca. 1. Moreover, the analysis reveals that the catalysts based on Ti-IMEC$_{NaOH}$, obtained with the use of NaOH during the neutralization step, contain a significant amount of sodium. This information was the incentive for the preparation of the titania-clay component by a modified procedure, using ammonia solution as the neutralizing agent. The resulting Ti-IMEC$_{NH3}$ derived composites do not contain Na, and maintain a Mn content comparable to the Ti-IMEC$_{NaOH}$ based counterparts. The titania-free composites have a similar content of Mn to the Ti-IMEC-based composites. The MnAl(st)/Na-Mt sample contains some sodium stemming from the clay component, while the MnAl(im)/CTA-Mt, obtained from the organoclay, is Na-free.

Table 1. XRF (X-ray fluorescence) determined chemical composition of the composites.

Sample	SiO_2 [wt%]	Al_2O_3 [wt%]	MgO [wt%]	TiO_2 [wt%]	MnO [wt%]	Fe_2O_3 [wt%]	Na_2O [wt%]
MnAl(im)/Ti-IMEC$_{NaOH}$-I	30.3	12.3	1.6	38.3	7.3	1.3	7.8
MnAl(im)/Ti-IMEC$_{NaOH}$-II	24.9	12.4	1.4	33.0	18.2	1.2	7.9
MnAl(im)/Ti-IMEC$_{NH3}$-I	33.7	13.2	1.8	41.5	7.2	1.5	-
MnAl(im)/Ti-IMEC$_{NH3}$-II	28.5	13.8	1.5	36.6	17.5	1.3	-
MnAl(im)/CTA-Mt	49.2	26.0	2.9	-	18.5	2.3	-
MnAl(st)/Na-Mt	47.7	25.1	2.7	-	19.8	2.3	2.0

XRD diagrams of composite catalysts calcined at 450 and 600 °C are collated in Figure 1. In addition, the pattern of MnAl(im) calcined at 450 and 600 °C is shown, to facilitate identification of reflections associated with the Mn-Al mixed oxide active phase. After calcination at 450 °C, the Mn-Al mixed oxide phase shows several very broad features, pointing to the formation of highly amorphized ε-MnO_2 (ref. code 30-0820) known to be catalyzed by the presence of aluminum [29], while after treatment at 650 °C, a clear set of peaks characteristic of well crystallined Mn_3O_4 (ref. code 24-0734) emerges. The XRD patterns of TiO_2-free MnAl(im)/CTA-Mt-450 and MnAl(im)/CTA-Mt-600 materials are relatively simple, as the only well resolved features can be identified as belonging to calcined montmorillonite and the quartz impurity. The position of the low intensity (001) reflection of montmorillonite corresponds to the interplanar distance $d_{001} \approx 1.0$ nm, the value characteristic of clay with thermally collapsed layers. In fully exfoliated clay with the house of cards structure this reflection should be absent, therefore its appearance indicates that partial restacking of exfoliated clay layers occurred. No reflection that might be attributed to a Mn-bearing oxide phase is visible, even after calcination at 600 °C. This result is at variance with the behavior of analogical composites based on Laponite, where upon calcination at 600 °C crystallization of Mn_3O_4 was observed [19]. The result suggests that the layers of synthetic Laponite offer a more efficient thermal barrier than those of montmorillonite, so that the heat evolved upon combustion of interlayer CTA cations in Laponite matrix is more likely to generate hotspots facilitating crystallization of Mn_3O_4. It has been proposed that one of the reasons for differences in thermal properties of organoclays based on Laponite, as opposed to montmorillonite, is that in the synthetic Laponite the silica sheets adhering to the organic matter are free of heteroatom impurities, while in montmorillonite the tetrahedral sheet properties are modified by the Al for Si substitution [30]. The MnAl(st)/Na-Mt-450 sample obtained by the inorganic route also displays only the features of thermally collapsed montmorillonite and quartz impurity, but in the MnAl(st)/Na-Mt-600 material additional reflections appear, pointing to the crystallization of the Mn_2O_3 phase.

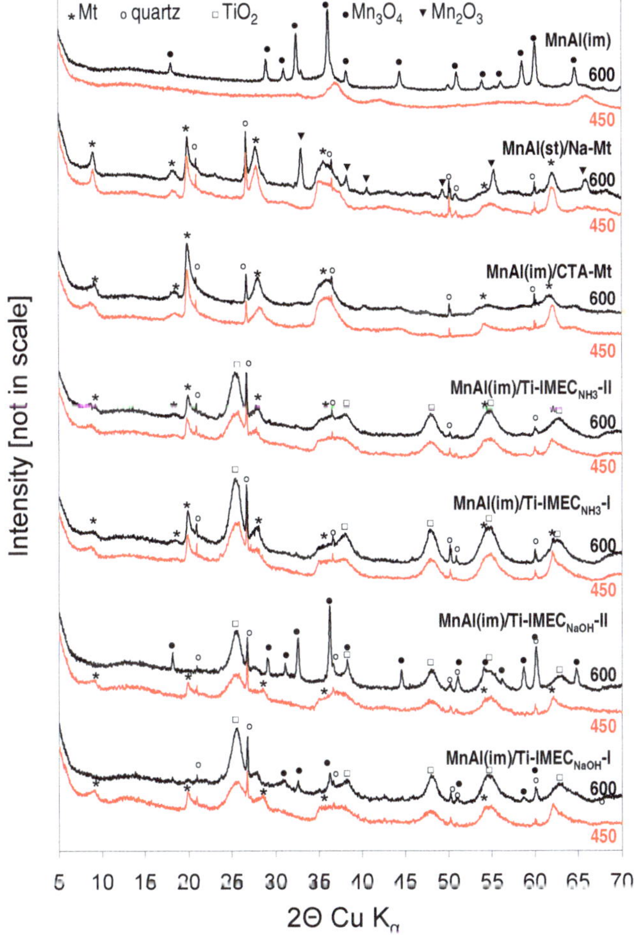

Figure 1. Powder X-ray diffraction (XRD) patterns of investigated composites and the MnAl(im) active phase calcined at 450 and 600 °C.

In all Ti-IMEC based samples the formation of the anatase modification of TiO_2 (ref. code 21-1272) is observed already after calcination at 450 °C (Figure 1). Titania crystallinity is improved upon calcination at 600 °C, where, according to Scherrer calculations, the coherently scattering domains of anatase reach ca. 6–7 nm. Analysis of the XRD patterns of Ti-IMEC based composites shows that in all samples calcined at 450 °C, whether prepared with NH_3 or NaOH, only features attributable to montmorillonite, anatase, and quartz impurity are visible, with no evidence of any Mn-containing oxide phase. Also, no crystalline Na-containing phase could be detected in samples obtained from

Ti-IMEC$_{NaOH}$. This shows that sodium either enters the clay interlayer, and/or becomes incorporated into anatase, and/or forms amorphous sodium titanate-like phase. The response of Ti-IMEC$_{NH3}$ and Ti-IMEC$_{NaOH}$ derived composites to thermal treatment at 600 °C is different. In the former the evidence of Mn-containing oxide crystallization is hard to find, while in MnAl(im)/Ti-IMEC$_{NaOH}$-I-600 and, especially, in MnAl(im)/Ti-IMEC$_{NaOH}$-II-600, sharp reflections of Mn$_3$O$_4$ appear. Moreover, in MnAl(im)/Ti-IMEC$_{NaOH}$-I-600 and in MnAl(im)/Ti-IMEC$_{NaOH}$-II-600, the features related to montmorillonite disappear. This points to the loss of the long range order in the clay component and shows that the Ti-IMEC$_{NaOH}$ based composites possess lower structural stability than the Ti-IMEC$_{NH3}$ derived ones. Both effects, i.e., the more facile degradation of clay and the ease of Mn$_3$O$_4$ crystallizatiom are likely to be due to the presence of sodium, whose compounds are known to act as fluxes enhancing solid state transformations [31,32]. It should be noted, however, that according to previous study [25], in the absence of MnAl(im) component no destruction of clay layers occurs in Ti-IMEC$_{NaOH}$ calcined at 600 °C, showing that the former must also be involved in the observed structural transformation. The estimated average crystal size of Mn$_3$O$_4$ in MnAl(im)/Ti-IMEC$_{NaOH}$-II-600 is ca. 55–60 nm.

N$_2$ adsorption/desorption isotherms obtained for the composite catalysts calcined at 450 and 600 °C are presented in Figure 2, and the textural parameters are gathered in Table 2.

Analysis of isotherms was carried out based on the recent IUPAC report [33]. All isotherms may be classified as Type II, and most of the accompanying hysteresis loops are of type H3 (Figure 2a,b,e,f). The strong upward swing of these isotherms is the result of unrestricted monolayer-multilayer adsorption up to high p/p$_0$. Such an isotherm shape indicates that the material contains both mesopores, which are responsible for the hysteresis, and macropores, which adsorb nitrogen in the relative pressure range of 0.98–1.00. In the case of catalysts derived from Ti-IMEC$_{NH3}$, the hysteresis loops are of a more complex shape. In MnAl(im)/Ti-IMEC$_{NH3}$-II-450 and MnAl(im)/Ti-IMEC$_{NH3}$-II-600 (Figure 2d) desorption branches display a two-step profile, encountered in materials containing a contribution from mesopores with partially plugged passages [34,35]. The desorption step at higher relative pressure reflects emptying of open mesopores, while the blocked pores remain filled until the relative pressure drops below p/p$_0$ ~0.5 and the cavitation of the nitrogen condensate leads to N$_2$ desorption from the confined areas. Such hysteresis loops are referred to as type H5. Samples MnAl(im)/Ti-IMEC$_{NH3}$-I-450 and MnAl(im)/Ti-IMEC$_{NH3}$-I-600 (Figure 2c) show also features of a two-step pore emptying, although less pronounced, and shifted to lower p/p$_0$. Therefore, the resulting hysteresis loops are intermediate between H3 and H5 type, indicating that pore blocking effects play a role also in these samples.

The isotherms of TiO$_2$-free MnAm(im)/CTA-Mt-450 and MnAm(im)/CTA-Mt-600 samples are of type II with H3 hysteresis loops (Figure 2e), and are very similar to those reported for analogical composites based on Laponite [19]. The exceptionally good thermal stability of the material, evidenced by the virtual lack of impact of the calcination temperature on the isotherm character and position, is another feature common with Laponite-derived composites [19]. The MnAm(st)/Na-Mt-450 and MnAm(st)/Na-Mt-600 composites, obtained by an inorganic route, have much lower adsorption capacity and poorer thermal stability than the MnAm(im)/CTA-Mt-450 and MnAm(im)/CTA-Mt-600 counterparts, as manifested by the position of their isotherms (Figure 2f) and the textural parameters given in Table 2. All titania-containing composites also show a downward shift of the isotherms upon increase of the calcination temperature from 450 to 600 °C. It may be presumed that thermal evolution of the titania component, visible in the XRD patterns, is a factor contributing to the observed changes of the composites' texture. The effect of textural shrinkage is particularly visible in the case of composites based on Ti-IMEC$_{NaOH}$ (Figure 2, Table 2), whose specific surface area after calcination at 600 °C is reduced by over 50%, to be compared with ca. 30% reduction observed for Ti-IMEC$_{NH3}$ derived samples. The fall of the specific surface is accompanied by an increase of the average pore diameter of composites calcined at 600 °C, which, again, is particularly strong for Ti-IMEC$_{NaOH}$ related composites (Table 2).

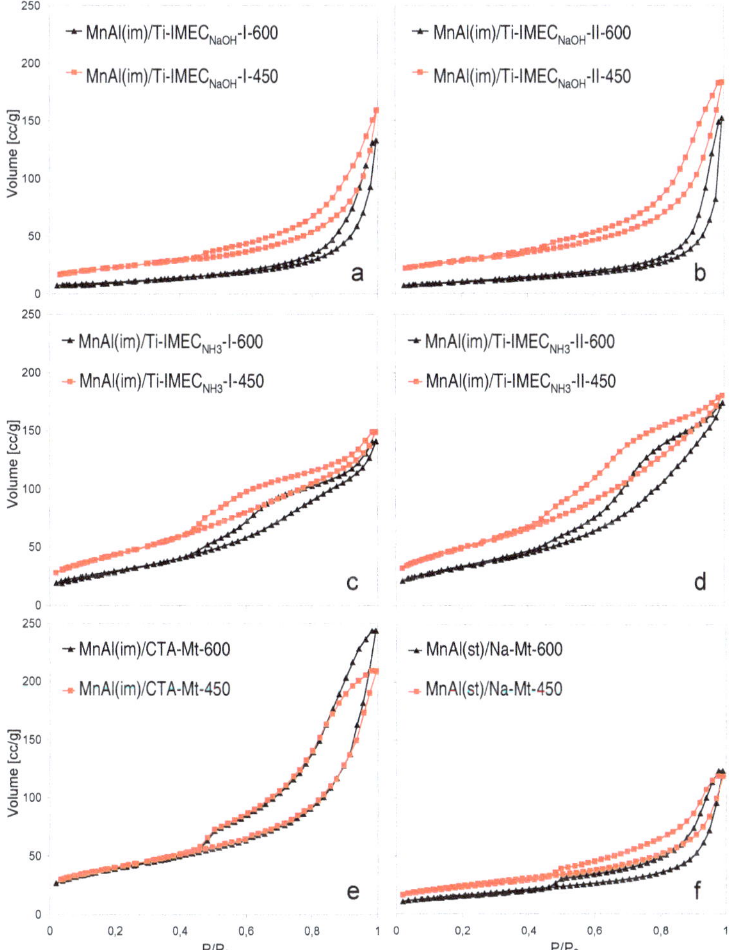

Figure 2. N_2 adsorption/desorption isotherms of investigated composites calcined at 450 and 600 °C, (**a**) MnAl(im)/Ti-IMEC$_{NaOH}$-I; (**b**) MnAl(im)/Ti-IMEC$_{NaOH}$-II; (**c**) MnAl(im)/Ti-IMEC$_{NH3}$-I; (**d**) MnAl(im)/Ti-IMEC$_{NH3}$-II; (**e**) MnAl(im)/CTA-Mt; (**f**) MnAl(st)/Na-Mt.

Evolution of PSD profiles in composites with comparable loading of the active phase upon increase of the calcination temperature also shows evident differences between the materials (Figure 3). Thus, the MnAl(im)/Ti-IMEC$_{NH3}$-II-450 sample has a narrow pore size distribution (PSD), shifted to slightly higher pore widths after calcination at 600 °C. In contrast, the PSD of MnAl(im)/Ti-IMEC$_{NaOH}$-II-450 is much broader and bimodal, while in MnAl(im)/Ti-IMEC$_{NaOH}$-II-600 most of the porosity below pore size 250 nm vanishes. The PSD of the titania-free MnAl(im)/CTA-Mt material is almost independent of the calcination temperature, which confirms that it is the recrystallization of TiO_2 nanoparticles that contributes to the temperature-induced textural changes in titania-bearing materials. However, the accelerated textural collapse of MnAl(im)/Ti-IMEC$_{NaOH}$ samples indicates that yet another factor influences the thermal evolution of these materials. Since the effect coincides with the XRD data pointing to the destruction of the clay lattice upon thermal treatment, the textural caving-in is attributed to the flux-like action of sodium present in these materials, causing local softening of the solid components and closure of the pore network.

Table 2. S^{BET}—specific surface area (in brackets % of the 450 °C specific surface retained at 600 °C), V_{tot}—total pore volume, D^{av}—average pore diameter, type of hysteresis loop, H/Mn—hydrogen consumption from TPR experiments, T_{50}—temperature of 50% conversion, T_{90}—temperature of 90% conversion, ΔT_{50}—difference between T_{50} after calcination at 600 and 450 °C, and ΔT_{90}—difference between T_{90} after calcination at 600 and 450 °C.

Sample	S_{BET} [m²/g]	V_{tot} [cm³/g]	D^{av} [nm]	Loop	H/Mn	T_{50} [°C]	T_{90} [°C]	ΔT_{50} [°C]	ΔT_{90} [°C]
MnAl(im)/Ti-IMEC$_{NaOH}$-I-450	81	0.25	12.25	H3	-	280	304	24	27
MnAl(im)/Ti-IMEC$_{NaOH}$-I-600	39 (48%)	0.20	20.72	H3	-	304	331		
MnAl(im)/Ti-IMEC$_{NaOH}$-II-450	90	0.28	12.53	H3	1.1	265	295	26	27
MnAl(im)/Ti-IMEC$_{NaOH}$-II-600	36 (40%)	0.24	25.94	H3	1.0	291	322		
MnAl(im)/Ti-IMEC$_{NH3}$-I-450	161	0.23	5.72	H3/H5	-	232	265	20	10
MnAl(im)/Ti-IMEC$_{NH3}$-I-600	108 (67%)	0.22	8.09	H3/H5	-	252	275		
MnAl(im)/Ti-IMEC$_{NH3}$-II-450	178	0.28	6.28	H5	1.6	206	239	21	8
MnAl(im)/Ti-IMEC$_{NH3}$-II-600	123 (69%)	0.26	8.58	H5	1.4	225	252		
MnAl(im)/CTA-Mt-450	132	0.33	10.08	H3	1.1	252	275	3	2
MnAl(im)/CTA-Mt-600	125 (95%)	0.36	11.39	H3	1.0	255	277		
MnAl(st)/Na-Mt-450	87	0.18	8.28	H3	1.5	259	292	20	17
MnAl(st)/Na-Mt-600	59 (68%)	0.19	12.88	H3	1.3	279	309		

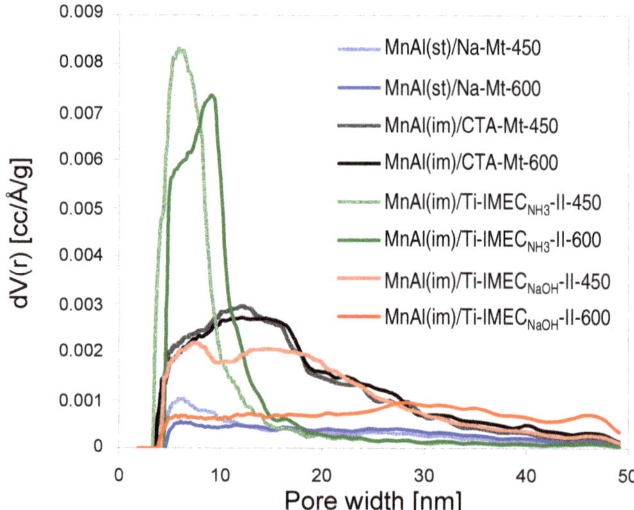

Figure 3. Pore size distribution profiles of investigated composites calcined at 450 and 600 °C.

High magnification SEM images of selected uncoated composites and components used for their synthesis are gathered in Figure 4. Thus, Figure 4a,b shows the morphology of the precipitate obtained from neutralized micellar TiO_2 precursor and of the as received MnAl(im), respectively. In both cases the powders are composed of fine, uniform particles formed under the influence of spatial constraints exerted by the limited volume of inverse micelles. Figure 4c shows the fluffy house of cards texture of exfoliated CTA-Mt deposited on the copper grid from dispersion in 1-hexanol. Figure 4d–f shows, respectively, the images of MnAl(im)/Ti-IMEC$_{NaOH}$-II-600, MnAl(im)/Ti-IMEC$_{NH3}$-II-600, and MnAl(im)/CTA-Mt-600 composite materials. It is visible that the appearance of both latter materials (Figure 4e,f) is determined by the loose, haphazard arrangement of platey clay particles, resembling that of CTA-Mt (Figure 4c). No obvious agglomerates of introduced nanoparticles of TiO_2 and/or Mn-Al mixed oxide are present. In contrast, in MnAl(im)/Ti-IMEC$_{NaOH}$-II-600 (Figure 4d) areas exist where the clay particles appear to fuse with the agglomerates of nanoparticles of oxide components. The image is consistent with the results of XRD and textural analyses indicating that Ti-IMEC$_{NaOH}$ are thermally less stable and upon calcination at 600 °C both the structure and the texture of the materials collapses. One may envisage that the growing structural disorder is associated with softening of the layers whose partial merger leads to strong modification of the materials texture. The occurrence of nanoparticle agglomerates may also be the reason for the appearance in MnAl(im)/Ti-IMEC$_{NaOH}$-II-600 of Mn_3O_4 with crystal size (55–60 nm) exceeding the thickness of a single primary MnAl(im) particle (Figure 4b).

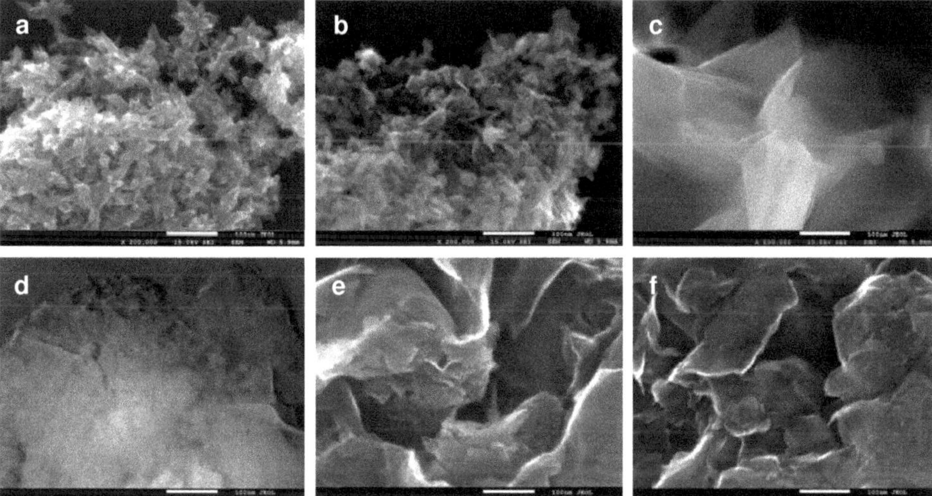

Figure 4. High resolution (HR) SEM images of (**a**) precipitate obtained from neutralized micellar TiO_2 precursor; (**b**) as received MnAl(im); (**c**) CTA-Mt exfoliated in 1-hexanol; (**d**) MnAl(im)/Ti-IMEC$_{NaOH}$-II-600; (**e**) MnAl(im)/Ti-IMEC$_{NH3}$-II-600; (**f**) MnAl(im)/CTA-Mt-600.

Reducibility of the catalysts, an important factor influencing the material's activity in total oxidation processes, was assessed by means of temperature programmed reduction (TPR) with hydrogen. Figure 5 shows the TPR profiles recorded for samples with similar Mn content, calcined at 450 or 600 °C. All major hydrogen consumption effects are due to the reduction of manganese-containing oxide phases. Of the other composite components, in the investigated temperature range only montmorillonite clay shows a very weak maximum around 600 °C, attributed to the reduction of structural iron (Figure 5). Transformation of manganese oxides upon reduction proceeds according to the general scheme: $MnO_2 \rightarrow Mn_2O_3 \rightarrow Mn_3O_4 \rightarrow MnO$ [36]. Inspection of Figure 5 shows that TPR curves differ significantly depending on the type of composites. In addition, in most cases the course of reduction depends visibly on the temperature of calcination, because MnOx oxides are known to lose lattice oxygen upon increasing the temperature of thermal treatment in air [37,38]. In most samples the TPR profiles are essentially bimodal. The low temperature maxima occurring below 400 °C are attributed to the reduction of Mn^{4+}, present in MnO_2-like species and/or as defects in oxygen-rich surface layers of other MnOx forms [19,36–39], to Mn^{3+} in Mn_2O_3-like stoichiometry. The subsequent reduction step of Mn_2O_3 to Mn_3O_4 is usually not distinguishable as a separate maximum, because it overlaps with the former effect [36]. The second maximum, occurring in the 450–520 °C temperature range, is assigned to the consecutive reduction of Mn_3O_4 to MnO. The situation in which the intensity of the first maximum prevails over the high temperature effects, as in the case of MnAl(im)/Ti-IMEC$_{NH3}$-II-450, MnAl(im)/Ti-IMEC$_{NH3}$-II-450 composites and the MnAl(st)/Na-Mt-450 sample prepared by the inorganic route, is indicative of a substantial content of MnO_2 [39]. This conclusion is corroborated by the H/Mn ratio found for these materials (Table 2). The theoretical hydrogen consumption for MnO_2 corresponds to H/Mn = 2, for Mn_2O_3 to H/Mn = 1 and for Mn_3O_4 to H/Mn ≈ 0.7. In the case of MnAl(im)/Ti-IMEC$_{NH3}$-II-450, MnAl(im)/Ti-IMEC$_{NH3}$-II-450, and MnAl(st)/Na-Mt-450 the observed H/Mn ratios equal 1.6, 1.4, and 1.5, respectively, which is meaningfully higher than the value expected for Mn_2O_3 reduction.

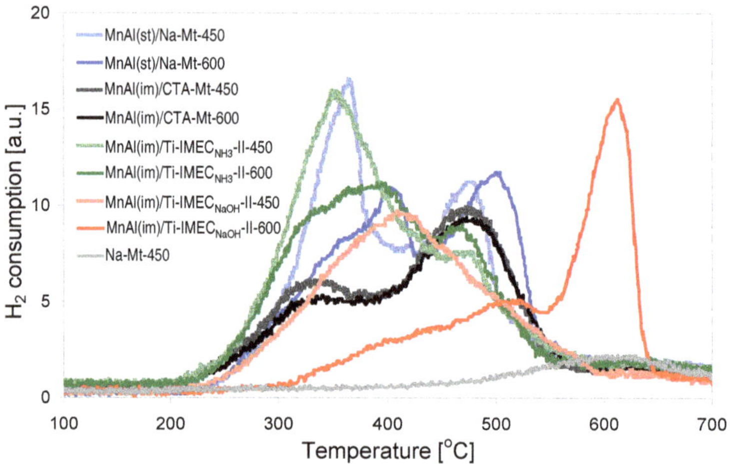

Figure 5. H_2 temperature programmed reduction (TPR) profiles of investigated composites calcined at 450 and 600 °C.

The H/Mn ratio is also quite high in the MnAl(st)/Na-Mt-600 sample, whose XRD pattern points to crystallization of Mn_2O_3. This means, that apart from the sesquioxide detected by XRD analysis, the material also contains Mn^{4+} species, which are responsible for the initial low temperature rise of the TPR profile, while the shift of the first maximum to higher temperature (around 400 °C) reflects the contribution from the reduction of the crystalline Mn_2O_3 component. The TPR profiles of the titania-free MnAl(im)/CTA-Mt-450 and MnAl(im)/CTA-Mt-600 samples are very similar, showing that the composite changes very little upon increase of the calcination temperature. The first maximum is lower than the high temperature one, indicating a lower content of Mn^{4+} species than in the MnAl(im)/Ti-IMEC$_{NH3}$-II composite which contains addition of titania, and in the MnAl(st)/Na-Mt sample prepared by the inorganic route. The H/Mn ratio found for the MnAl(im)/CTA-Mt samples is around 1 (Table 2), which confirms the lower average oxidation state of Mn in these materials and points to the beneficial role of using Ti-IMEC$_{NH3}$ for enhancing the content of easily reducible Mn species formed upon calcination of the MnAl(im) component. The composites obtained from Ti-IMEC$_{NaOH}$ possess very different TPR characteristics. In the MnAl(im)/Ti-IMEC$_{NaOH}$-II-450 sample a single, broad maximum centered around 400 °C is observed, clearly being an envelope of several overlapping effects, which indicates a heterogeneity of Mn species valencies and environments, of average oxidation state ca. +3, as shown by the H/Mn ratio of 1.1. After calcination at 600 °C a substantial change in redox properties is observed, clearly associated with the evolution of the phase composition and destruction of the clay layers. The characteristic feature of the MnAl(im)/Ti-IMEC$_{NaOH}$-II-600 material is a remarkable shift of TPR maxima towards higher temperature. The H/Mn ratio of 1 corresponds formally to Mn_2O_3-like stoichiometry. Comparison with TPR curves of other composites shows a virtual lack of easily reducible component, attributed in other materials to the reduction of Mn^{4+} species. In view of the XRD detected presence of Mn_3O_4 phase in this sample, the maximum around 500 °C may be attributed to the reduction of this phase to MnO, while the broad effect preceding this maximum shows that some more reducible MnO_x species, probably in less ordered environment, are present as well. Bearing in mind the XRD evidence of clay lattice destruction, accompanied by textural collapse, the strong second maximum, located at a much higher temperature than any of the effects found in other composites, suggests that it is due to manganese contained in another, amorphous phase, formed during solid component softening with the flux effect of sodium. To comply with the overall hydrogen consumption, manganese is expected to be present in this phase as Mn^{3+}. It should be mentioned that the detrimental effect of sodium addition on the reducibility and textural properties of the MnO_x/TiO_2 system has been reported previously [40].

Thus, on the basis of the physico-chemical characterization one may conclude that in the case of TiO_2-free composites, the use of an organic route yields MnAl(im)/CTA-Mt material of much better developed textural properties and excellent thermal stability, but characterized by a lower average oxidation state of Mn than in the MnAl(st)/Na-Mt sample obtained by the inorganic preparative procedure. The effect of titania addition depends whether NaOH or NH_3 solutions are used in the Ti-IMEC neutralization step. The catalysts based on Ti-IMEC$_{NH3}$ are characterized by the presence of a pretty uniform and thermally stable mesoporous network, and high abundance of easily reducible Mn species. In contrast, the composites derived from Ti-IMEC$_{NaOH}$ are the least reducible of all studied composites and possess more heterogeneous porosity, which due to the presence of sodium, suffers thermal collapse upon increase of the calcination temperature from 450 to 600 °C. The schematic models of the composites structure after treatment at 600 °C are shown in Figure 6.

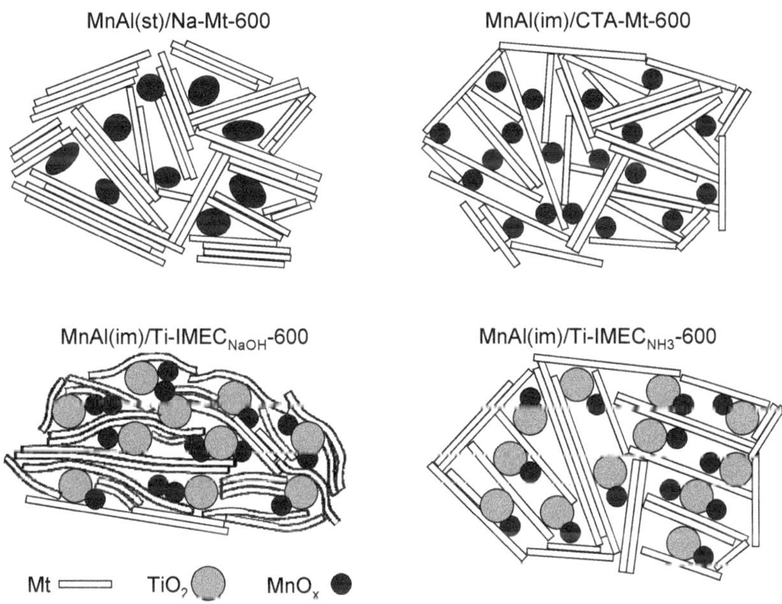

Figure 6. Schematic models of investigated composites calcined at 600 °C.

3.2. Catalytic Testing

Catalytic activity of the composites was tested in the reaction of total combustion of toluene, frequently used as a model volatile organic compound. Figure 7a,b shows the toluene ignition curves of the composites calcined at 450 and 600 °C, respectively. Table 2 provides information on the temperatures at which 50% (T_{50}) and 90% (T_{90}) toluene conversion is reached for each of the catalysts.

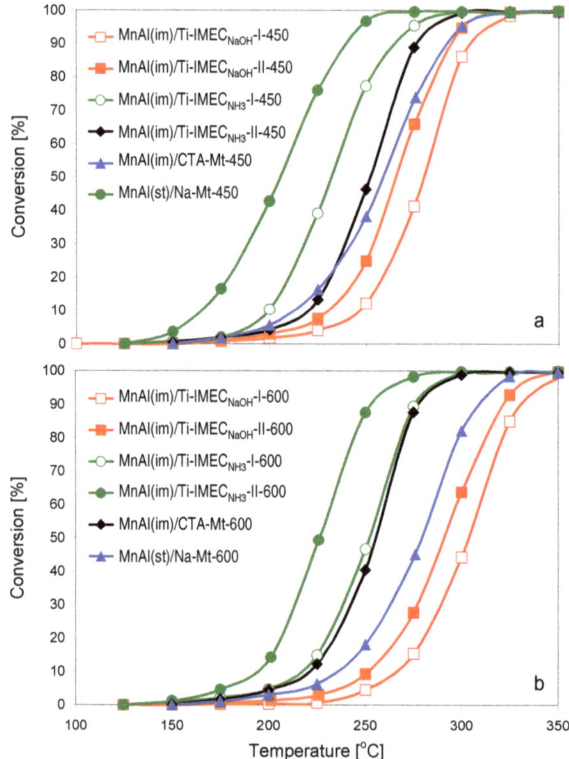

Figure 7. Ignition curves for toluene combustion over investigated composites calcined at (**a**) 450 °C; (**b**) 600 °C.

Analysis of Figure 7 shows that all investigated catalysts are characterized by very high activity, but differences are observed depending on the nature of the composite. Thus, it is evident that the MnAl(im)/CTA-Mt catalyst prepared by the organic route, performs better than the MnAl(st)/Na-Mt one, obtained by the inorganic procedure with conventionally synthesized Ht. This is especially visible when comparing performance after calcination at 600 °C, which, as revealed by the physico-chemical characterization has a negligible effect on the MnAl(im)/CTA-Mt material, while the MnAl(st)/Na-Mt is much more strongly affected, both in terms of structure, texture, and reducibility. In consequence, the MnAl(im)/CTA-Mt composite is a catalyst whose activity is practically independent on the calcination temperature, and the values of $\Delta_{50} = T_{50}{}^{600} - T_{50}{}^{450}$ and $\Delta_{90} = T_{90}{}^{600} - T_{90}{}^{450}$ are only 3 °C and 2 °C, respectively (Table 2). The relevant values for the MnAl(st)/Na-Mt catalyst equal 20 and 17 °C, which clearly shows the advantage of trapping MnAl(im) Ht nanoparticles between the organoclay layers, over the more conventional synthesis route. The enhanced stability of MnO_x phase generated from precursor embedded in organoclay is attributed to the fact that during calcination the combustion of interlayer CTA cations causes an additional increase of local temperature. The maximum of the exothermic effect associated with the combustion of organic matter appears around 300 °C [19], i.e., lower than any of the employed calcination temperatures. In consequence, the MnO_x component evolving during calcination experiences a temperature higher than the nominal one, which hardens the formed oxide against thermal degradation.

For both temperatures of calcination the highest activity is observed for the Ti-IMEC$_{NH3}$ based composites, which show spectacular combustion performance, the ones with higher MnAl(im) loading, MnAl(im)/Ti-IMEC$_{NH3}$-II-450 and MnAl(im)/Ti-IMEC$_{NH3}$-II-600, reaching 90% toluene conversion at 239 and 252 °C, respectively. This can be explained in terms of the catalysts' physico-chemical characteristics, which show that the materials combine two features important for the catalytic reaction: good and relatively stable textural properties and a high contribution of easily reducible manganese component. It should be noted that according to the TPR data, incorporation of titania into the composite improves the reducibility of the MnO$_x$ phase formed upon calcination of the MnAl(im) component. This is in accordance with previous reports indicating that titania-supported MnO$_x$ catalysts exhibited lower reduction temperatures than the unsupported forms [41,42]. However, the performance of Ti-IMEC$_{NaOH}$-based catalysts is inferior not only to the Ti-IMEC$_{NH3}$-derived ones but also to the titania-free catalysts. Comparison of the data in Table 2 shows that not only are the absolute values of T$_{50}$ and T$_{90}$ significantly higher for catalysts obtained from Ti-IMEC$_{NaOH}$ than for those prepared with Ti-IMEC$_{NH3}$, but the former are characterized by higher $\Delta_{50} = T_{50}^{600} - T_{50}^{450}$ and $\Delta_{90} = T_{90}^{600} - T_{90}^{450}$ differences, which shows that they are more strongly affected by the increase of the temperature of calcination. This concerns in particular the $\Delta_{90} = T_{90}^{600} - T_{90}^{450}$ values, which are much lower for MnAl(im)/Ti-IMEC$_{NH3}$-I and MnAl(im)/Ti-IMEC$_{NH3}$-II, due to the steeper profiles of their light-off curves observed at 600 °C. The difference between MnAl(im)/Ti-IMEC$_{NaOH}$-II-450 and MnAl(st)/Na-Mt-450, containing a comparable amount of Mn, is not very large. Since both catalysts display similar specific surface areas, the lower activity of the former appears to be related chiefly to the lower average oxidation state of the Mn species and their higher reduction temperature. After calcination at 600 °C, the difference becomes more pronounced, which reflects the effect of the collapsed porous network in MnAl(im)/Ti-IMEC$_{NaOH}$-II-600, and further deterioration of the active phase reducibility. Thus, the presence of sodium in the composites obtained from Ti-IMEC$_{NaOH}$, shown to affect their structural, textural and redox properties, has also a profound adverse effect on their catalytic properties. In contrast, the use of Ti-IMEC$_{NH3}$ in combination with MnAl(im) component leads to composite catalysts of excellent combustion activity. Comparison with the literature data shows that in toluene combustion these materials perform considerably better than other previously described clay-transition metal oxide composite catalysts, e.g. Cu,Ce/Zr-PILC (T$_{90}$ = 300 °C) [10], Fe-PILC (T$_{90}$ = 380 °C) [15], or Fe/Ti-PILC (T$_{90}$ = 347 °C) [18].

4. Conclusions

Physico-chemical characterization of novel clay/TiO$_2$/MnAl-mixed oxide composites, in which non-clay components are obtained by an inverse microemulsion method, reveals profound differences in structural, textural, and redox properties of the materials, depending on the presence of the TiO$_2$ component, the manner of titania nanoparticles preparation, and the temperature of calcination. Thus, the trapping of MnAl Ht nanoparticles, obtained from inverse microemulsion, between the organoclay layers, yields, after calcination, composite material of much better textural properties and with excellent thermal stability, but a lesser average oxidation state of Mn than the analogical material prepared from conventionally synthesized MnAl Ht and the sodium form of montmorillonite. The effect of insertion of MnAl Ht nanoparticles into clay, previously loaded with microemulsion containing TiO$_2$ precursor, depends whether NaOH or NH$_3$ solutions are used at the neutralization step. The materials obtained with the use of ammonia are characterized by the presence of a pretty uniform and thermally stable mesoporous network, coupled with a high abundance of easily reducible Mn species. In particular, their reducibility is better than that of composites without titania. In contrast, the composites derived from NaOH treated TiO$_2$ precursor are significantly less reducible and possess more heterogeneous porosity, which, due to the presence of sodium, suffers thermal collapse upon increase of the calcination temperature from 450 to 600 °C. Catalytic tests of toluene combustion reveal that the clay/TiO$_2$/MnAl-mixed oxide composites prepared with the use of ammonia show excellent activity, the composites obtained from MnAl Ht nanoparticles trapped between the organoclay layers are less active, but display spectacular thermal stability, while the clay/TiO_2/MnAl-mixed oxide materials obtained with aid of NaOH are least active. The observed patterns of catalytic activity bear a direct relation to the structural, textural, and redox properties of the materials.

Acknowledgments: This work was supported financially by the Polish National Science Center (NCN) grant OPUS 2013/09/B/ST5/00983. K.B. acknowledges financial contribution of AGH University of Science and Technology project 11.11.140.319.

Author Contributions: B.D.N., A.M.-Z., M.R., E.B., W.R., A.G. and M.W.-B. performed the experiments and analyzed the data, E.M.S. and K.B. conceived and designed the experiments, interpreted the results and wrote the paper.

Conflicts of Interest: The authors declare no conflict of interest. The founding sponsors had no role in the design of the study; in the collection, analyses, or interpretation of data; in the writing of the manuscript, and in the decision to publish the results.

References

1. Everaert, K.; Baeyens, J. Catalytic combustion of volatile organic compounds. *J. Hazard. Mater. B* **2004**, *109*, 113–139. [CrossRef] [PubMed]

2. Storaro, L.; Ganzerla, R.; Lenarda, M.; Zanoni, R.; Jiménez López, A.; Olivera-Pastor, P.; Rodróguez Castellón, E. Catalytic behavior of chromia and chromium-doped alumina pillared clay materials for the vapor phase deep oxidation of chlorinated hydrocarbons. *J. Mol. Catal. A* **1997**, *115*, 329–338. [CrossRef]

3. Bahranowski, K.; Bielańska, E.; Janik, R.; Machej, T.; Serwicka, E.M. LDH-derived catalysts for complete oxidation of volatile organic compounds. *Clay Miner.* **1999**, *34*, 67–77. [CrossRef]

4. Kovanda, F.; Jiratova, K.; Rymes, J. Characterization of activated Cu/Mg/Al hydrotalcites and their catalytic activity in toluene combustion. *Appl. Clay Sci.* **2001**, *18*, 71–80. [CrossRef]

5. Gandía, L.M.; Vicente, M.A.; Gil, A. Complete oxidation of acetone over manganese oxide catalysts supported on alumina- and zirconia-pillared clays. *Appl. Catal. B* **2002**, *38*, 295–307. [CrossRef]

6. Serwicka, E.M.; Bahranowski, K. Environmental catalysis by tailored materials derived from layered minerals. *Catal. Today* **2004**, *90*, 85–92. [CrossRef]

7. Dula, R.; Janik, R.; Machej, T.; Stoch, J.; Grabowski, R.; Serwicka, E.M. Mn-containing catalytic materials for the total combustion of toluene: The role of Mn localisation in the structure of LDH precursor. *Catal. Today* **2007**, *119*, 327–331. [CrossRef]

8. Lamonier, J.F.; Boutoundou, A.B.; Gennequin, C.; Pérez-Zurita, M.J.; Siffert, S.; Aboukais, A. Catalytic removal of toluene in air over Co-Mn-Al nano-oxides synthesized by hydrotalcite route. *Catal. Lett.* **2007**, *118*, 165–172. [CrossRef]

9. Mishra, T.; Mohapatra, P.; Parida, K.M. Synthesis, characterisation and catalytic evaluation of iron-manganese mixed oxide pillared clay for VOC decomposition reaction. *Appl. Catal. B* **2008**, *79*, 279–285. [CrossRef]

10. Chen, M.; Fan, L.; Qi, L.; Luo, X.; Zhou, R.; Zheng, X. The catalytic combustion of VOCs over copper catalysts supported on cerium-modified and zirconium-pillared montmorillonite. *Catal. Commun.* **2009**, *10*, 838–841. [CrossRef]

11. Palacio, L.A.; Velásquez, J.; Echavarría, A.; Faro, A.; Ribeiro, F.R.; Ribeiro, M.F. Total oxidation of toluene over calcined trimetallic hydrotalcites type catalysts. *J. Hazard. Mater.* **2010**, *177*, 407–413. [CrossRef] [PubMed]

12. Aguilera, D.A.; Perez, A.; Molina, R.; Moreno, S. Cu-Mn and Co-Mn catalysts synthesized from hydrotalcites and their use in the oxidation of VOCs. *Appl. Catal. B* **2011**, *104*, 144–150. [CrossRef]

13. Kovanda, F.; Jirátová, K. Supported layered double hydroxide-related mixed oxides and their application in the total oxidation of volatile organic compounds. *Appl. Clay Sci.* **2011**, *53*, 305–316. [CrossRef]

14. Genty, E.; Cousin, R.; Capelle, S.; Gennequin, C.; Siffert, S. Catalytic oxidation of toluene and CO over nanocatalysts derived from hydrotalcite-like compounds ($X_6^{2+}Al_2^{3+}$): Effect of the bivalent cation. *Eur. J. Inorg. Chem.* **2012**, 2802–2811. [CrossRef]

15. Li, D.; Li, C.S.; Suzuki, K. Catalytic oxidation of VOCs over Al- and Fe-pillared montmorillonite. *Appl. Clay Sci.* **2013**, *77–78*, 56–60. [CrossRef]

16. Machej, T.; Serwicka, E.M.; Zimowska, M.; Dula, R.; Michalik-Zym, A.; Napruszewska, B.; Rojek, W.; Socha, R. Cu/Mn-based mixed oxides derived from hydrotalcite-like precursors as catalysts for methane combustion. *Appl. Catal. A* **2014**, *474*, 87–94. [CrossRef]

17. Michalik-Zym, A.; Dula, R.; Duraczyńska, D.; Kryściak-Czerwenka, J.; Machej, T.; Socha, R.P.; Włodarczyk, W.; Gaweł, A.; Matusik, J.; Bahranowski, K.; et al. Active, selective and robust Pd and/or Cr catalysts supported on Ti-, Zr- or [Ti,Zr]-pillared montmorillonites for destruction of chlorinated volatile organic compounds. *Appl. Catal. B* **2015**, *174*, 293–307. [CrossRef]

18. Liang, X.; Qi, F.; Liu, P.; Wei, G.; Su, X.; Ma, L.; He, H.; Lin, X.; Xi, Y.; Zhu, J.; et al. Performance of Ti-pillared montmorillonite supported Fe catalysts for toluene oxidation: The effect of Fe on catalytic activity. *Appl. Clay Sci.* **2016**, *132–133*, 96–104. [CrossRef]

19. Napruszewska, B.D.; Michalik-Zym, A.; Dula, R.; Bielanska, E.; Rojek, W.; Machej, T.; Socha, R.P.; Lityńska-Dobrzyńska, L.; Bahranowski, K.; Serwicka, E.M. Composites derived from exfoliated Laponite and Mn-Al hydrotalcite prepared in inverse microemulsion: A new strategy for design of robust VOCs combustion catalysts. *Appl. Catal. B* **2017**, *211*, 46–56. [CrossRef]

20. Krishnamoorthy, S.; Rivas, J.A.; Amiridis, M.D. Catalytic Oxidation of 1,2-Dichlorobenzene over Supported Transition Metal Oxides. *J. Catal.* **2000**, *193*, 264–272. [CrossRef]

21. Liu, Y.; Luo, M.; Wei, Z.; Xin, Q.; Ying, P.; Li, C. Catalytic oxidation of chlorobenzene on supported manganese oxide catalysts. *Appl. Catal. B* **2001**, *29*, 61–67. [CrossRef]

22. Morales, M.R.; Barbero, B.P.; Lopez, T.; Moreno, A.; Cadús, L.E. Evaluation and characterization of Mn-Cu mixed oxide catalysts supported on TiO_2 and ZrO_2 for ethanol total oxidation. *Fuel* **2009**, *88*, 2122–2129. [CrossRef]

23. Doggali, P.; Teraoka, Y.; Mungse, P.; Shah, I.K.; Rayalua, S.; Labhsetwar, N. Combustion of volatile organic compounds over Cu-Mn based mixed oxide type catalysts supported on mesoporous Al_2O_3, TiO_2 and ZrO_2. *J. Mol. Catal. A* **2012**, *358*, 23–30. [CrossRef]

24. Zhu, X.; Zhang, S.; Yu, X.; Zhu, X.; Zheng, C.; Gao, X.; Luo, Z.; Cen, K. Controllable synthesis of hierarchical MnO_X/TiO_2 composite nanofibers for complete oxidation of low-concentration acetone. *J. Hazard. Mater.* **2017**, *337*, 105–114. [CrossRef] [PubMed]

25. Bahranowski, K.; Gaweł, A.; Klimek, A.; Michalik-Zym, A.; Napruszewska, B.D.; Nattich-Rak, M.; Rogowska, M.; Serwicka, E.M. Influence of purification method of Na-montmorillonite on textural properties of clay mineral composites with TiO_2 nanoparticles. *Appl. Clay Sci.* **2017**, *140*, 75–80. [CrossRef]

26. Bellezza, F.; Cipiciani, A.; Costantino, U.; Nocchetti, M.; Posati, T. Hydrotalcite-like nanocrystals from water-in-oil microemulsions. *Eur. J. Inorg. Chem.* **2009**, *18*, 2603–2611. [CrossRef]

27. Venugopal, B.R.; Sen, S.; Shivakumara, C.; Rajamathi, M. Delamination of surfactant intercalated smectites in alcohols: Effect of chain length of the solvent. *Appl. Clay Sci.* **2006**, *32*, 141–146. [CrossRef]

28. Sterte, J. Synthesis and properties of titanium oxide cross-linked montmorillonite. *Clays Clay Miner.* **1986**, *34*, 658–664. [CrossRef]

29. Ryu, W.H.; Han, D.W.; Kim, W.K.; Kwon, H.S. Facile route to control the surface morphologies of 3D hierarchical MnO_2 and its Al self-doping phenomenon. *J. Nanopart. Res.* **2011**, *13*, 4777–4784. [CrossRef]

30. Ovadyahu, D.; Lapides, I.; Yariv, S. Thermal analysis of tributylammonium montmorillonite and Laponite. *J. Therm. Anal. Calorim.* **2007**, *87*, 125–134. [CrossRef]

31. Perrichon, V.; Durupty, M.C. Thermal stability of alkali metals deposited on oxide supports and their influence on the surface area of the support. *Appl. Catal.* **1988**, *42*, 217–227. [CrossRef]

32. Bondioli, F.; Corradi, A.B.; Manfredini, T.; Leonelli, C.; Bertoncello, R. Nonconventional synthesis of praseodymium-doped ceria by flux method. *Chem. Mater.* **1999**, *12*, 324–330. [CrossRef]

33. Thommes, M.; Kaneko, K.; Neimerk, A.V.; Olivier, J.P.; Rodriguez-Reinozo, F.; Rouquerol, J.; Sing, K.S.W. Physisorption of gases with special reference to the evaluation of surface area and pore size distribution (IUPAC Technical Report). *Pure Appl. Chem.* **2015**, *879*, 1051–1069. [CrossRef]

34. Van Der Voort, P.; Ravikovitch, P.I.; De Jong, K.P.; Neimark, A.V.; Janssen, A.H.; Benjelloun, M.; Van Bavel, E.; Cool, P.; Weckhuysen, B.M.; Vansant, E.F. Plugged hexagonal templated silica. A unique micro- and mesoporous composite material with internal silica nanocapsules. *Chem. Commun.* **2002**, 1010–1011. [CrossRef]

35. Kruk, M.; Jaroniec, M.; Joo, S.H.; Ryoo, R. Characterization of regular and plugged SBA-15 silicas by using adsorption and inverse carbon replication and explanation of the plug formation mechanism. *J. Phys. Chem.* **2003**, *107*, 2205–2213. [CrossRef]
36. Kapteijn, F.; Singoredjo, L.; Andreini, A. Activity and selectivity of pure manganese oxides in the selective catalytic reduction of nitric oxide with ammonia. *Appl. Catal. B* **1994**, *3*, 173–189. [CrossRef]
37. Stobbe, E.R.; de Boer, B.A.; Geus, J.W. The reduction and oxidation behaviour of manganese oxides. *Catal. Today* **1999**, *47*, 161–167. [CrossRef]
38. Gil, A.; Gandía, L.M.; Korili, S.A. Effect of the temperature of calcination on the catalytic performance of manganese- and samarium-manganese-based oxides in the complete oxidation of acetone. *Appl. Catal. A* **2004**, *274*, 229–235. [CrossRef]
39. Fu, X.; Feng, J.; Wang, H.; Ng, K.M. Manganese oxide hollow structures with different phases: Synthesis, characterization and catalytic application. *Catal. Commun.* **2009**, *10*, 1844–1848. [CrossRef]
40. Fang, D.; He, F.; Xie, J.L.; Dong, P.P. Influence of sodium on MnO_X/TiO_2 catalysts for SCR of NO with NH_3 at low temperature. *Mater. Res. Innov.* **2014**, *18*, S45–S49. [CrossRef]
41. Hu, J.; Chu, W.; Shi, L. Effects of carrier and Mn loading on supported manganese oxide catalysts for catalytic combustion of methane. *J. Nat. Gas Chem.* **2008**, *17*, 159–164. [CrossRef]
42. Fang, D.; Xie, J.; Hu, H.; Yang, H.; He, F.; Fu, Z. Identification of MnO_X species and Mn valence states in MnO_X/TiO_2 catalysts for low temperature SCR. *Chem. Eng. J.* **2015**, *271*, 23–30. [CrossRef]

Article

Microstructural Modification and Characterization of Sericite

Yu Liang, Hao Ding *, Sijia Sun and Ying Chen

School of Materials Science and Technology, China University of Geosciences, No. 29 Xueyuan Road, Haidian District, Beijing 100083, China; liangyuaadd@gmail.com (Y.L.); 1012122105@cugb.edu.cn (S.S.); chenying@cugb.edu.cn (Y.C.)
* Correspondence: dinghao113@126.com or dinghao@cugb.edu.cn

Received: 24 August 2017; Accepted: 7 October 2017; Published: 16 October 2017

Abstract: Activated sericite was prepared by thermal modification, acid activation and sodium modification, and it was characterized by X-ray diffraction (XRD) analysis, differential scanning calorimetry (DSC), N_2 adsorption test, thermo-gravimetric analysis (TGA), nuclear magnetic resonance (NMR), and scanning electron microscope (SEM). The results indicated that the crystallinity of raw sericite decreased after thermal modification; the pores with sizes between 5 nm to 10 nm of thermal-modified sericite have collapsed and the surface area increased after thermal modification. The dissolving-out amount of Al^{3+} reached ca. 31 mg/g in the optimal processing conditions during acid activation; cation exchange capacity (CEC) of acid-treated sericite increased to 56.37 mmol/100 g meq/g after sodium modification compared with that of raw sericite (7.42 mmol/100 g). The activated sericite is a promising matrix for clay-polymer nanocomposites.

Keywords: sericite; thermal modification; acid activation; sodium modification

1. Introduction

In recent years, great attention has been paid to clay-polymer nanocomposites due to their extraordinary properties. Compared with pure polymer, this category of composites usually exhibits higher moduli [1–3], larger strength and heat resistance [4], smaller gas permeability [5–7], better fire retardancy [8,9], higher ionic conductivity [10], and increased biodegradability of biodegradable polymers [11–13]. Clay-polymer nanocomposites are widely used in a range of key areas, such as aerospace, automobile, appliances, and electronics [14–16]. These properties depend heavily on the structure of nanocomposites, which is determined by the physical properties of clay mineral, such as its cation exchange capacity (CEC) which is used to quantify the excess negative charge of layered silicates and their capability to exchange ions. CEC is highly dependent on the nature of the isomorphous substitutions in the tetrahedral and octahedral layers.

The common layered silicates used for preparation of clay-polymer nanocomposites are 2:1 type (montmorillonite, vermiculite, and mica) and 1:1 type (kaolinite); the former is used much more frequently. Montmorillonite [17–22] and vermiculite [23–29] have been mostly investigated as the matrix materials for clay-polymer nanocomposites because of their swelling behavior and ion exchange properties.

Although sericite belongs to 2:1 clay minerals, it does not swell in water and has almost no ion exchange capacity. It can hardly be intercalated because it has a high layer charge density close to 1.0 equivalent per $O_{10}(OH)_2$, which produces pretty strong electrostatic force [30]. This layer charge stems mainly from the substitution of Al^{3+} for Si^{4+} in tetrahedral sheet. Sericite is very fined squamous-structured muscovite, as one kind of mica family. It holds the advantages of high moduli, stable chemical property, high electrical insulation and good ultraviolet ray resistance [31–34]. Therefore, it is necessary to explore the preparation of an expandable sericite with relatively high CEC.

The purpose of activation is to permanently reduce the layer charge of sericite and obtain a number of exchangeable cations. The physical and chemical modifications have long been used to activate clay and clay minerals, such as acid activation and thermal treatment. Salt modification, mechanical grinding [35,36] and swelling by the decomposition of hydrogen peroxide [37] are also usually employed. Poncelet and del Rey-Perez-Caballero [25,38] permanently reduced the global negative charge of the mineral layers by the combination of calcination and acid activation, and the resulting product (activated vermiculite/phlogopite mica) was successfully used as matrix in micro porous 18 Å Al-pillared nanocomposites. They found that Al^{3+} in tetrahedral sheet could be partly dissolved out by the combination process and yet had an unremarkable effect on the structure of octahedral sheet. Furthermore, a well-swelled sericite with 80% exchangeable K^+ and a CEC of 110 mmol/100 g was obtained by Shih and Shen [32] by thermal modification and Li-hydrothermal treatment.

In this study, activated sericite was prepared by thermal modification, acid activation, and sodium modification. The X-ray diffraction (XRD) analysis, N_2 adsorption test, nuclear magnetic resonance (NMR) and scanning electron microscope (SEM) were used to elucidate the effect of this process. From our study, it can be seen that thermal modification can reduce the layer charge and crystallinity of sericite. Acid activation can dissolve both the octahedral and tetrahedral Al^{3+} out and reduce its layer charge and sodium modification can finally improve the CEC value of sericite. Therefore, the whole modification process can make sericite more suitable for polymer-clay nanocomposites.

2. Materials and Methods

2.1. Materials

The raw sericite (S_0) was obtained from Anhui province, China. Its mean size is about 10 μm. CEC of S_0 is 7.42 mmol/100 g (0.07 meq/g). The quantitative analysis of the raw material showed that the purity of raw sericite is 93.2%, with 6.8% of quartz. Ruling out the influence of quartz, the chemical composition of S_0 is listed in Table 1. The chemical formula is $(K_{0.79}Na_{0.11}Ca_{0.01})(Al_{1.64}Ti_{0.02}Fe_{0.18}Mg_{0.24})(Al_{0.92}Si_{3.08})O_{10}(OH)_2$.

Table 1. Chemical composition of the original sericite.

Composition	SiO_2	Al_2O_3	Fe_2O_3	TiO_2	K_2O	Na_2O	CaO	MgO	SO_3	L.O.I	Total
Content (mass %)	45.71	28.32	3.04	0.35	8.09	0.71	0.10	2.12	0.075	4.47	99.555

2.2. Preparation

A certain amount of raw sericite (S_0) was put into Al_2O_3 crucibles and heated between 500 and 1000 °C in muffle for 1 to 3 h and cooled to room temperature naturally. After that, the thermally-treated product (S_1) was stirred with different kinds and concentrations of acid between 60 to 95 °C in thermostatic water bath for 4 h. The acid-treated product (S_2) was washed, filtrated, and dried at 80 °C. In this study, experiment term was based on an orthogonal term array experimental design (OA (9, 3^4)) where the following four variables were analyzed: the kinds of acid (factor A), acid concentration (factor B), reaction temperature (factor C) and reaction time (factor D). Finally, sodium chloride was added to react with S_2 in round bottomed flask. The sodium modified product (S_3) was obtained by mixing, washing, centrifuging, and drying at 80 °C. The orthogonal experiment term method (OA (9, 3^3)) was used to find three optimal parameters: concentration of Na^+ (factor A), reaction temperature (factor B) and reaction time (factor C).

2.3. Characterization

The X-ray diffraction patterns were obtained on a Rigaku Rotaflex X-ray powder diffractometer (Rigaku, Tokyo, Japan), employing Cu Kα radiation, 40 kV, 100 mA. The X-ray diffraction (XRD) patterns in the 2θ range from 3°–70° were collected at 4°/min. Simultaneous collection of DSC and

TGA signals was carried out using a SDT Q600 analyzer (TA, New Castle, DE, USA) under air flow and heated from room temperature to 1100 °C at 10 °C/min. The BET surface area of the samples was determined by N_2 adsorption by using NOVA4000 equipment (Quantachrome, Boynton Beach, FL, USA). Prior to N_2 adsorption, the samples were evacuated at 473 K under vacuum for 4 h. The pore size distribution was calculated using the BJH method. ^{27}Al NMR spectrum (130.327 Hz) was recorded on a Bruker Avance III spectrometer (Bruker, Karlsruher, Germany). The dwell time is 0.01 s and the rotational speed is 6000 rpm.

3. Results and Discussion

3.1. Thermal Modification

A slight weight loss is observed in the TG curve at low temperature (Figure 1), which is attributed to absorbed surface water. There is a mass loss in the TG curve at 3% between 670 and 841 °C according to the DSC curve peak at the same temperature, which indicates that the hydroxyl groups were lost with increasing temperature during thermal modification.

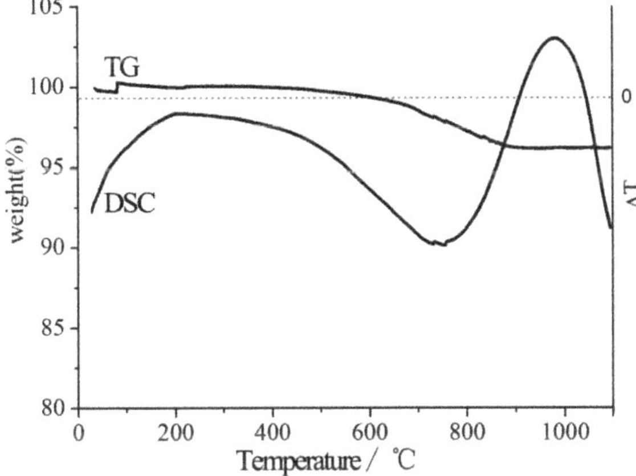

Figure 1. TG and DSC curves of S_0.

The XRD patterns of sericite which were activated at different temperatures are shown in Figure 2. The intensities of major reflections decreased gradually as the temperature increased and finally almost disappeared at 1100 °C, which suggested that the mica-type phase persisted when the temperature was lower than 900 °C. The phase transformation would occur and its crystal integrity would be destroyed gradually when heated at 1000 °C. The lattice activated degree can be judged by lattice distortion level. The relationship between lattice distortion level, crystal size, full width at half maximum (FWHM) of reflections and diffraction angle can be calculated by Scherrer's equation:

$$Bcos\ \theta = K\lambda/D + 4\Delta dsin\ \theta/d \tag{1}$$

where B is FWHM; θ is diffraction angle; K is the form factor, which is close to 1; λ is the wavelength of X-ray; D is the crystal lattice size; d is the distance between crystal planes; Δd is the average deviation between the distance of reflecting planes under study and the mean value d. $4\Delta d/d$ shows the level of lattice distortion. The larger the value, the higher the distortion level, and vice versa. In this part, Bcos θ and sin θ of different products which reflect the lattice distortion level were calculated from XRD data. The slope $(4\Delta d/d)$ and intercept $(k\lambda/d)$ were obtained by linear fitting, using sin θ as X axis and Bcos θ as Y axis. The obtained values were summarized in Table 2.

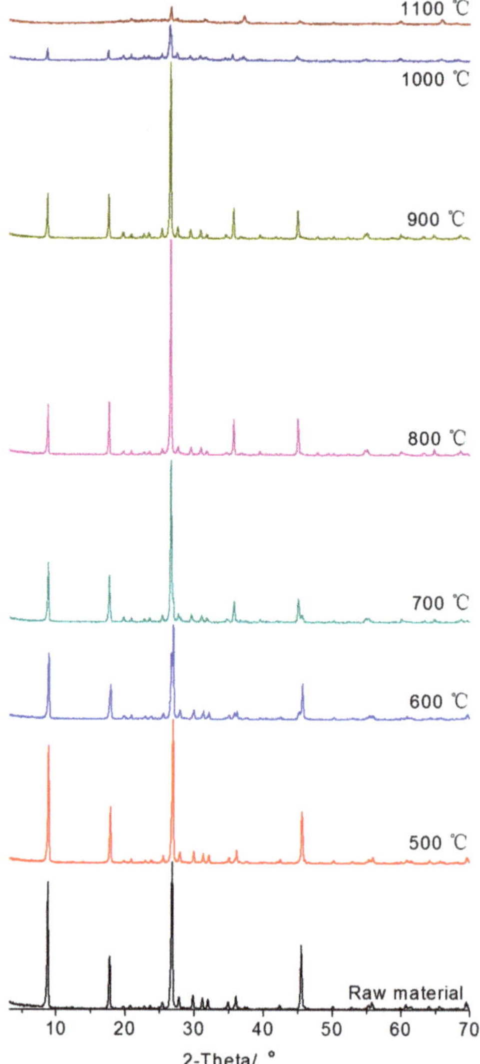

Figure 2. XRD patterns of sericite after heating at different temperatures.

Table 2. Relation between Bcosθ and sinθ of sericite activated at different temperatures in the (002) reflection.

Temperature (°C)	Raw Material	500	600	700	800	900	1000
$k\lambda/D$	0.2146	0.2259	0.3059	0.3797	0.2937	0.2128	0.1938
$4\Delta d/d$	0.0118	−0.0381	−0.1363	−0.1631	−0.1674	−0.1372	−0.0407

As shown in Table 2, the value of $4\Delta d/d$ of raw sericite is 0.0118 (close to 0), which is a proof of little lattice distortion. As the temperature increased, the absolute value of $4\Delta d/d$ of the thermally-treated product increased gradually, reaching the largest value at 800 °C. The line's slope starts to decrease at 900 °C, follows by an even larger decrease at 1000 °C. The results above demonstrate that the best activated temperature is 800 °C (with the same holding time). The loss of crystallinity is evaluated using the FWHM index (Table 3). Heating at 800 °C got the largest FWHM, which is a sign of the most lattice defects and distortion.

The raw sericite samples were heated at 800 °C and preserved for 1 h, 2 h, and 3 h. The XRD patterns are shown in Figure 3. The values of kλ/D and 4Δd/d extracted from the XRD data are shown in Table 4. The absolute value of 4Δd/d of sericite with a holding time of 1 h is much higher than that of raw material. When the holding time increased to 2 h or more, the slope decreased. The potential reason is that the increasing holding time make the unobvious preferential orientation of flakes caused by lattice distortion increase. Therefore, the best preservation time is 1 h.

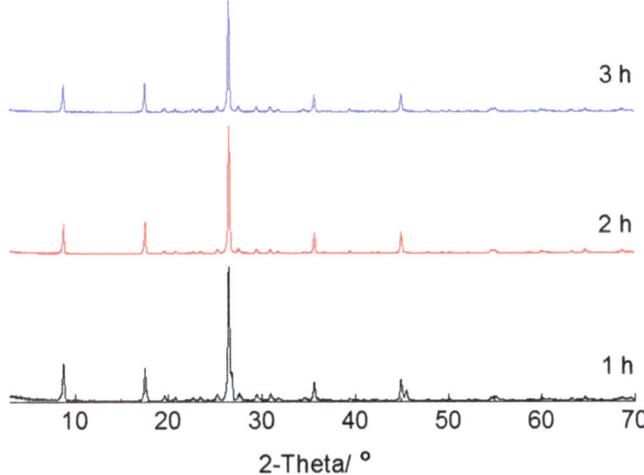

Figure 3. XRD patterns of sericite with different holding time at 800 °C.

Table 3. FWHM index for raw material and thermal-modified products in the (002) reflection.

Samples	FWHM (°)
S_0	0.137
S_1 (500 °C)	0.138
S_1 (600 °C)	0.168
S_1 (700 °C)	0.173
S_1 (800 °C)	0.174
S_1 (900 °C)	0.141
S_1 (1000 °C)	0.132
S_1 (1100 °C)	-

Table 4. Relation between Bcosθ and sinθ of sericite with different holding time.

Holding Time (h)	Raw Material	1	2	3
kλ/D	0.2146	0.4287	0.2937	0.2277
4Δd/d	0.0118	−0.2848	−0.1674	−0.1026

N$_2$ physisorption measurements have also been performed on both S_0 and S_1 (Figure 4). It can be seen that after thermal modification, the pores with sizes between 5 nm to 10 nm of thermally-modified sericite have collapsed. The surface area of S_0 is 14.653 m^2/g, while the surface area of S_1 is 16.579 m^2/g, which means thermal modification increased the activity of sericite.

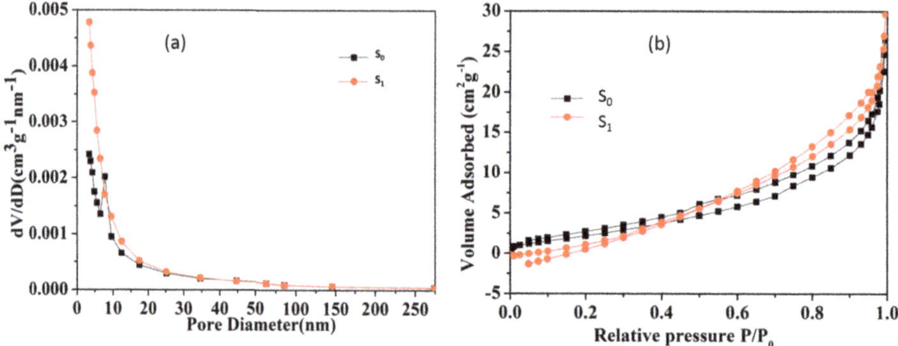

Figure 4. (a) Pore diameter distribution of raw sericite (S_0) and thermal-activated sericite (S_1); (b) the isotherm of N_2 adsorption-desorption on sericite before and after thermal modification.

3.2. Acid Activation

According to the research of Poncelet and del Rey-Perez-Caballero [25] on the activation of vermiculite and phlogopite, the combination of acid treatment and heat treatment was employed to modify the microstructure of sericite. As a result, Al^{3+} was dissolved out and the negative layer charge was reduced, which enables sericite to take on an ion exchange capacity. The results of acid activation were evaluated by dissolving-out an amount of Al^{3+}. The larger the dissolving-out amount of Al^{3+}, the better the effect of acid activation.

The main four factors, kinds of acid (factor A), acid concentration (factor B), reaction temperature (factor C), and reaction time (factor D) were researched and each control parameter has three experimental levels (Table 5) [39,40].

Table 5. Design and results of the orthogonal experiment of acid treatment of sericite [a].

Trial No.	Factors				Results
	Kinds of Acid *A*	Acid Concentration *B* (mol/L)	Reaction Temperature *C* (°C)	Reaction Time *D* (h)	Dissolving-Out Amount of Al^{3+} (mg/g)
1	HNO_3	1	60	1	4.2
2	HNO_3	3	80	3	12.8
3	HNO_3	5	95	5	31.0
4	H_2SO_4	1	80	5	9.9
5	H_2SO_4	3	95	1	14.0
6	H_2SO_4	5	60	3	8.2
7	HCl	1	95	3	11.1
8	HCl	3	60	5	8.5
9	HCl	5	80	1	9.9
$K_{1,j}$	48.0	25.2	20.9	28.1	-
$K_{2,j}$	32.1	35.3	32.6	32.1	-
$K_{3,j}$	29.5	49.1	56.1	49.4	-
$k_{1,j}$	16.0	8.4	7.0	9.4	-
$k_{2,j}$	10.7	11.8	10.9	10.7	-
$k_{3,j}$	9.8	16.4	18.7	16.5	-
R_j	6.2	8.0	11.7	7.1	-

[a]: K_{ij} is defined as the sum of the evaluation indexes of all levels (i, i = 1, 2, 3) in each factor (j, j = A, B, C, D) and k_{ij} (mean value of K_{ij}) is used to determine the optimal level and the optimal combination of factors. The optimal level for each factor could be obtained when k_{ij} is the largest; R_j is defined as the range between the maximum and minimum value of k_{ij} and is used for evaluating the importance of the factors.

The optimal values of different factors determined with reference to Table 5 are as follows: nitric acid, 5 mol/L, 95 °C, 5 h. In addition, the factors' levels of significance are as follows: reaction temperature > acid concentration > reaction time > type of acid.

Figure 5 shows the single effect of each factor on acid activation. A higher reaction temperature helped to dissolve Al^{3+} out. When temperature was low, the reaction system could not obtain enough power, so only a small amount of Al^{3+} dissolved out. The dissolving-out amount of Al^{3+} increased with the acid concentration, which means the higher the H^+ concentration, the better the result. However, when the acid concentration is ultrahigh, the layered structure would, conceivably, be seriously destroyed, which is not good for activation of sericite. Additionally, nitric acid is more effective on acid activation than the other two, although the kind of acid is not the most significant factor.

NMR analysis was done after acid activation. The range of chemical shift (δ) of Al is 450 ppm. Generally, δ of octahedral Al (Al_o) species and tetrahedral Al (Al_t) species is -10 to 10 ppm and 50–70 ppm, respectively. Therefore, ^{27}Al NMR is employed to distinguish the two kinds of Al in clay. As shown in Figure 6, δ of Al_t and Al_o of S_0 was 71.4 (spinning sidebands were 118 and 25, respectively), and 4.0 (sidebands were 50 and -42, respectively), both of which were similar to theoretical values. The counterparts of S_2 turned to be 67.5 and 4.0, respectively.

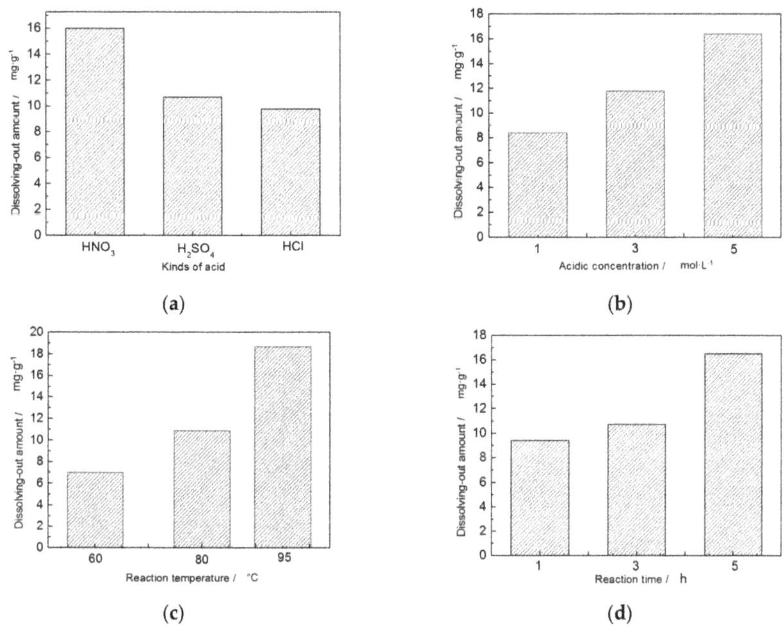

Figure 5. Effect of (**a**) kinds of acid; (**b**) acid concentration; (**c**) reaction temperature; and (**d**) reaction time on dissolving-out amount of Al^{3+}

The sharp peaks in ^{27}Al NMR is usually the sign of short range order in Al, while broad peaks signal short range disorder. The peaks of S_0 in ^{27}Al NMR spectrum are sharper than those of S_2, which indicate that the layered structure become more disordered. The peak intensities of Al_t and Al_o of S_2 decrease by 34% and 32%, respectively, as compared with those of S_0, which suggest that Al_t and Al_o are both dissolved out.

The relative content of Al_t in S_2 increased, and yet the peak width increased, which signaled an uneven distribution of Al_t. On the contrary, the relative content of Al_o decreased, and yet the peak width decreased, which is a sign of even distribution of Al_o. This phenomenon can be explained by the decrease of layer charges that leads to a higher order degree of Al_o [41]. The ratio of Al_t to Al_o decreased after acid treatment from 6.25:10 to 5.82:10, which means acid-treated sericite is more suitable for the ion exchanges in the next step.

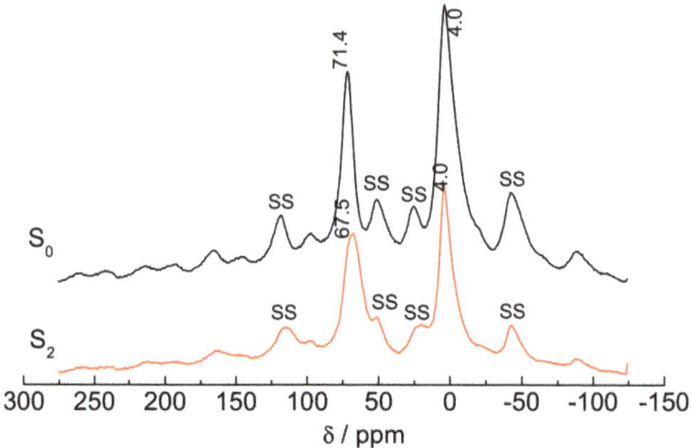

Figure 6. ^{27}Al NMR spectrums of S_0 and S_2 (SS means "spinning sidebands").

3.3. Sodium Modification

The results of sodium modification are evaluated by CEC. The larger the CEC value, the better the sodium modification result. Detailed sodium modification conditions are listed in Table 6. It can be seen that the optimal sodium modification conditions are as follows: supersaturated solution of sodium chloride, 95 °C, 3 h. The factors' levels of significance are as follows: Na^+ concentration > reaction temperature > reaction time. The single effect of each factor on sodium modification is shown in Figure 7, which indicates that higher concentration of Na^+ and higher reaction temperature are of great benefit to the CEC of S_3.

The interlayer potassium cation twelve coordinates with two aspectant hexagonal holes created by the Si/Al tetrahedral sheet, and is able to fit the two holes very tightly between the layers. Therefore, potassium cation and two adjacent tetrahedron sheets are bonded together closely by the electrostatic attraction. Consequently, it is hard for K^+ to be exchanged by Na^+. However, the combination of thermal modification and acid activation made the exchange possible, which was due to the activation of lattice. Na^+ has the superiority of the smaller hydrated ionic radius and lower hydrated energy compared with those of K^+. The higher reaction osmotic pressure was applied by the higher concentration of Na^+, and higher reaction temperature made ions turn to be more active. Therefore, higher concentration of Na^+ and reaction temperature would benefit sodium modification. Additionally, the exchange of Na^+ for K^+ was in the state of dynamic equilibrium. Therefore, longer reaction time has no effect on the CEC of S_3. Compared our study with the study of Shih, we used different treatment methods (we used Na-hydrothermal treatment while they used Li-hydrothermal treatment) at different temperatures (we used 60–95 °C while they used 90–270 °C). This is why our final CEC value (56.37 mmol/100 g) is lower than theirs (110 mmol/100 g).

The XRD patterns of raw material and activated products (S_1, S_2, and S_3) prepared at the optimal conditions are shown in Figure 8, and the loss of crystallinity is evaluated using the FWHM index (Table 7). The decrease of reflection intensities of S_1 was caused by the removal of the hydroxyl water of raw material corresponding to the increase of FWHM of S_1. After acid activation, the reflection intensities and crystallinity of S_2 further decreased, which was a sign of more lattice defects and larger lattice distortion. Compared with the pattern of S_2, the interlayer space of S_3 decreased slightly, which was due to the exchange of Na^+ for K^+ between layers. However, sodium modification led to a better crystallinity of S_3 than that of S_2. This may be due to the fact that Na^+ balanced the change of layer charge of S_2 caused by acid activation, and the crystal structure of S_2 was repaired to some extent.

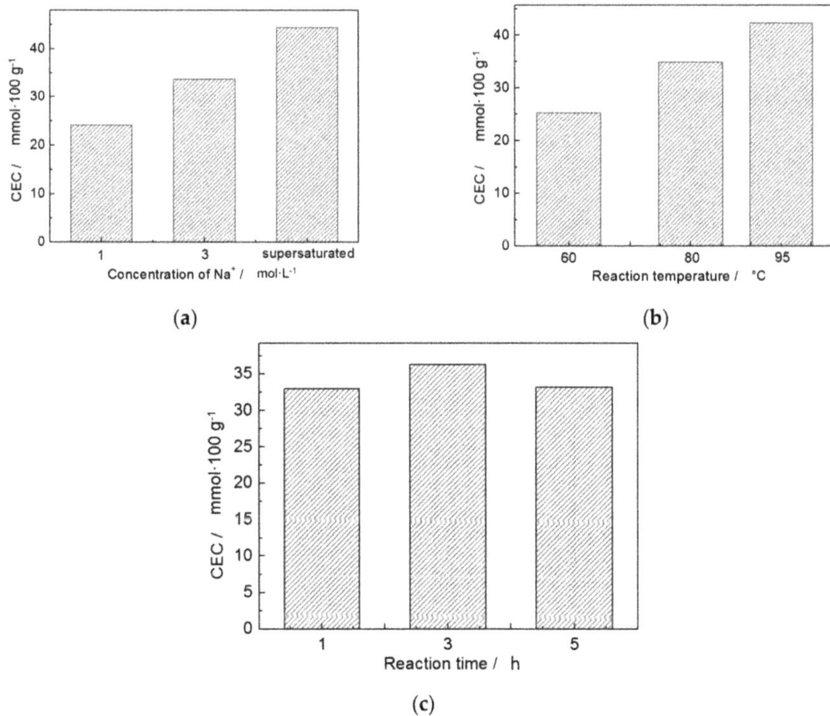

Figure 7. Effect of (**a**) concentration of Na^+; (**b**) reaction temperature; and (**c**) reaction time on CEC.

Table 6. Design and results of the orthogonal experiment of sodium modification of sericite.

Trial No.	Factors			Results
	Concentration of Na^+ A (mol/L)	Reaction Temperature B (°C)	Reaction Time C (h)	CEC (mmol/100 g)
1	1	60	1	15.76
2	1	80	3	27.24
3	1	95	5	29.70
4	3	60	3	25.36
5	3	80	5	35.24
6	3	95	1	40.76
7	Supersaturated	60	5	34.62
8	Supersaturated	80	1	42.34
9	Supersaturated	95	3	56.37
$K_{1,j}$	72.70	75.24	90.06	
$K_{2,j}$	101.36	104.82	108.97	-
$K_{3,j}$	133.33	126.83	99.56	-
$k_{1,j}$	24.23	25.25	32.95	-
$k_{2,j}$	33.79	34.94	36.32	-
$k_{3,j}$	44.44	42.28	32.19	-
R_j	20.21	17.03	4.13	-

Table 7. FWHM index for raw material and activate products in the (002) reflection.

Samples	FWHM (°)
S_0	0.208
S_1	0.226
S_2	0.720
S_3	0.452

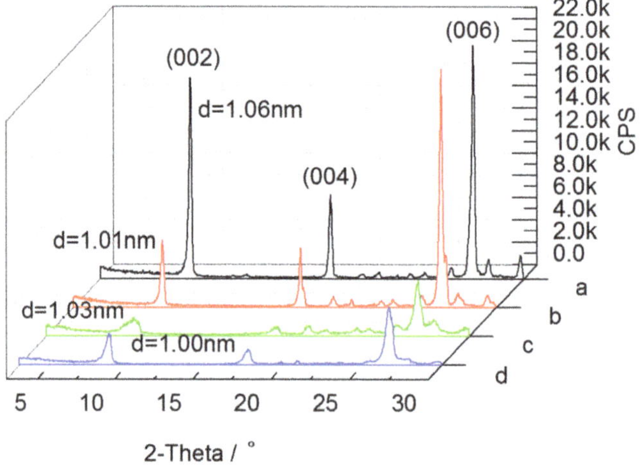

Figure 8. XRD patterns of (**a**) S_0; (**b**) S_1; (**c**) S_2; and (**d**) S_3.

From SEM images, it can be seen that S_0 has smooth surfaces, sharp fringed flakes, and uniform particle size (Figure 9). The SEM images of S_1, S_2, and S_3 clearly indicate that the mica-type phase of sericite persists while the particle surfaces become rougher.

Figure 9. SEM images of (**a**) S_0; (**b**) S_1; (**c**) S_2; and (**d**) S_3.

4. Conclusions

Activated sericite was prepared by thermal modification, acid activation, and sodium modification. The final product can be prepared by heating at 800 °C for 1 h, reacting with 5 mol/L nitric acid at 95 °C for 5 h and mixing with supersaturated solution of sodium chloride at 95 °C for 3 h. After modification, the mica-type phase persisted while its crystallinity decreased. The CEC of the final product can be enlarged from 7.42 mmol/100 g to 56.37 mmol/100 g meq/g. The activated sericite is much more suitable than raw sericite to prepare polymer-clay nanocomposites.

Author Contributions: Yu Liang and Hao Ding conceived and designed the experiments; Yu Liang performed the experiments and analyzed data; Yu wrote the paper. Sijia Sun and Ying Chen helped the experiments and analyzed data.

Conflicts of Interest: The authors declare no conflict of interest.

References

1. Kojima, Y.; Usuki, A.; Kawasumi, M.; Okada, A.; Fukushima, Y.; Kurauchi, T.; Kamigaito, O. Mechanical properties of nylon 6-nlay hybrid. *J. Mater. Res.* **1993**, *8*, 1185–1189.
2. Lebaron, P.C.; Wang, Z.; Pinnavaia, T.J. Polymer-layered silicate nanocomposites: An overview. *Appl. Clay Sci.* **1999**, *15*, 11–29.
3. Vaia, R.A.; Price, G.; Ruth, P.N.; Nguyen, H.T.; Lichtenhan, J. Polymer/layered silicate nanocomposites as high performance ablative materials. *Appl. Clay Sci.* **1999**, *15*, 67–92.
4. Giannelis, E.P. Polymer-layered silicate nanocomposites: Synthesis, properties and applications. *Appl. Organomet. Chem.* **1998**, *12*, 675–680.
5. Xu, R.; Manias, E.; Snyder, A.J.; Runt, J. New biomedical poly(urethane urea)—Layered silicate nanocomposites. *Macromolecules* **2001**, *34*, 337–339.
6. Messersmith, P.; Giannelis, E. Synthesis and barrier properties of poly(ε-caprolactone)—layered silicate nanocomposites. *J. Polym. Sci. A Polym. Chem.* **1995**, *33*, 1047–1057.
7. Akin, O.; Tihminlioglu, F. Effects of organo-modified clay addition and temperature on the water vapor barrier properties of polyhydroxy butyrate homo and copolymer nanocomposite films for packaging applications. *J. Polym. Environ.* **2017**, *25*, 1–12. [CrossRef]
8. Nguyen, Q.; Ngo, T.; Tran, P.; Mendis, P.; Bhattacharyya, D. Influences of clay and manufacturing on fire resistance of organoclay/thermoset nanocomposites. *Compos. Part A* **2015**, *74*, 26–37.
9. Gilman, J.; Jackson, C.; Morgan, A.; Harris, R.; Manias, E.; Giannelis, E.; Wuthenow, M.; Hilton, D.; Phillips, S. Flammability properties of polymer-layered-silicate nanocomposites. Polypropylene and polystyrene nanocomposites. *Chem. Mater.* **2000**, *12*, 1866–1873.
10. Aranda, P.; Mosqueda, Y.; Perez-Cappe, E.; Ruiz-Hitzky, E. Electrical characterization of poly(ethylene oxide)—clay nanocomposites prepared by microwave irradiation. *J. Polym. Sci. B Polym. Phys.* **2003**, *41*, 3249–3263.
11. Ray, S.S.; Yamada, K.; Okamoto, M.; Ueda, K. Polylactide-layered silicate nanocomposite: A novel biodegradable material. *Nano Lett.* **2002**, *2*, 1093–1096.
12. Kaur, M.; Datta, M. Synthesis and characterization of biodegradable clay-polymer nanocomposites for oral sustained release of anti-inflammatory drug. *Eur. Chem. Bull.* **2013**, *2*, 670–678.
13. Lilichenko, N.; Maksimov, R.D.; Zicans, J.; Meri, R.M.; Plume, E. A biodegradable polymer nanocomposite: Mechanical and barrier properties. *Mech. Compos. Mater.* **2008**, *44*, 45–56.
14. Pavlidou, S.; Papaspyrides, C.D. A review on polymer-layered silicate nanocomposites. *Prog. Polym. Sci.* **2008**, *22*, 1119–1198.
15. Sahoo, S.; Manjaiah, K.M.; Datta, S.C.; Shabeer, T.P.A., Kumar, J. Kinetics of metribuzin release from bentonite-polymer composites in water. *J. Environ. Sci. Health Part B Pestic. Food Contam. Agric. Wastes* **2014**, *49*, 591–600.
16. El-Hamshary, H.; Selim, A.I.; Salahuddin, N.A.; Mandour, H.S. Clay-polymer nanocomposite-supported brominating agent. *Clays Clay Miner.* **2015**, *63*, 328–336.
17. Choi, H.; Kim, S.; Hyun, Y.; Jhon, M. Preparation and rheological characteristics of solvent-cast poly(ethylene oxide)/montmorillonite nanocomposites. *Macromol. Rapid Commun.* **2001**, *22*, 320–325.

18. Hackett, E.; Manias, E.; Giannelis, E.P. Molecular dynamics simulations of organically modified layered silicates. *J. Chem. Phys.* **1998**, *108*, 7410–7415.

19. Strawhecker, K.E.; Manias, E. Structure and properties of poly(vinyl alcohol)/Na^+ montmorillonite nanocomposites. *Chem. Mater.* **2000**, *12*, 2943–2949.

20. Yano, K.; Usuki, A.; Okada, A.; Kuranchi, T.; Kamigaito, O. Synthesis and properties of polyimide clay hybrid. *J. Polym. Sci. A Polym. Chem.* **1993**, *31*, 2493–2498.

21. Ginzburg, V.; Balazs, A. Calculating phase diagrams for nanocomposites: The effect of adding end-functionalized chains to polymer/clay mixtures. *Adv. Mater.* **2000**, *12*, 1805–1809.

22. Manias, E.; Chen, H.; Krishnamoorti, R.; Genzer, J.; Kramer, E.J.; Giannelis, E.P. Intercalation kinetics of long polymers in 2 nm confinements. *Macromolecules* **2000**, *33*, 7955–7966.

23. Wang, L.; Chen, Z.; Wang, X.; Yan, S.; Wang, J.; Fan, Y. Preparations of organo-vermiculite with large interlayer space by hot solution and ball milling methods: A comparative study. *Appl. Clay Sci.* **2011**, *51*, 151–157.

24. Fonseca, M.G.D.; Wanderley, A.F.; Souea, K.; Arakaki, L.N.H.; Espinola, J.G.P. Interaction of aliphatic diamines with vermiculite in aqueous solution. *Appl. Clay Sci.* **2006**, *32*, 94–98.

25. Rey-Perez-Caballero, F.D.; Poncelet, G. Preparation and characterization of microporous 18 angstrom Al-pillared structures from natural phlogopite micas. *Microporous Mesoporous Mater.* **2000**, *41*, 169–181.

26. Williams-Daryn, S.; Thomas, R.K. The intercalation of a vermiculite by cationic surfactants and its subsequent swelling with organic solvents. *J. Colloid Interface Sci.* **2002**, *255*, 303–311.

27. Tjong, S.C.; Meng, Y.Z. Preparation and characterization of melt-compounded polyethylene/vermiculite nanocomposites. *J. Polym. Sci. B Polym. Phys.* **2003**, *41*, 1476–1484.

28. Tjong, S.C.; Meng, Y.Z.; Xu, Y. Structure and properties of polyamide-6/vermiculite nanocomposites prepared by direct melt compounding. *J. Polym. Sci. B Polym. Phys.* **2002**, *40*, 2860–2870.

29. Xu, J.; Meng, Y.Z.; Li, R.K.Y.; Rajulu, A.V. Preparation and properties of poly(vinyl alcohol)-vermiculite nanocomposites. *J. Polym. Sci. B Polym. Phys.* **2003**, *41*, 749–755.

30. Yu, X.; Ram, B.; Jiang, X. Parameter setting in a bio-inspired model for dynamic flexible job shop scheduling with sequence-dependent setups. *Eur. J. Ind. Eng.* **2007**, *1*, 182–199.

31. Ding, H.; Xu, X.; Liang, N.; Wang, Y. Preparation sericite nanoflakes by exfoliation of wet ultrafine grinding. *Adv. Mater. Res.* **2011**, *178*, 242–247.

32. Shih, Y.-J.; Shen, Y.-H. Swelling of sericite by LiNO(3)-hydrothermal treatment. *Appl. Clay Sci.* **2009**, *43*, 282–288.

33. Kim, J.-O.; Lee, S.-M.; Jeon, C. Adsorption characteristics of sericite for cesium ions from an aqueous solution. *Chem. Eng. Res. Des.* **2014**, *92*, 368–374.

34. Vaia, R.A.; Teukolsky, R.K.; Giannelis, E.P. Interlayer structure and molecular environment of alkylammonium layered silicates. *Chem. Mater.* **1994**, *6*, 1017–1022.

35. Valášková, M.; Barabaszová, K.; Hundáková, M.; Ritz, M.; Plevová, E. Effects of brief milling and acid treatment on two ordered and disordered kaolinite structures. *Appl. Clay Sci.* **2011**, *54*, 70–76.

36. Caseri, W.R.; Shelden, R.A.; Suter, U.W. Preparation of muscovite with ultrahigh specific surface-area by chemical cleavage. *Colloid Polym.Sci.* **1992**, *270*, 392–398.

37. Obut, A.; Girgin, I. Hydrogen peroxide exfoliation of vermiculite and phlogopite. *Miner. Eng.* **2002**, *15*, 683–687.

38. Rey-Perez-Caballero, F.D.; Poncelet, G. Microporous 18 angstrom Al-pilared vermiculites: Preparation and characterization. *Microporous Mesoporous Mater.* **2000**, *37*, 313–327.

39. Teng, H.; Zhu, G.; Huang, P.; Liu, P. Example analysis of orthogonal experimental design. *Pharm. Care Res.* **2008**, *1*, 75–76.

40. He, Z.; Liu, Y.; Chen, L.; Cao, M.; Xia, J. Orthogonal design-direct analysis for PCR optimization. *Bull. Hunan Med. Univ.* **1998**, *4*, 76–77.

41. Gates, W.P.; Komadel, P.; Madejova, J.; Bujdak, J.; Stucki, J.W.; Kirkpatrick, R.J. Electronic and structural properties of reduced-charge montmorillonites. *Appl. Clay Sci.* **2000**, *16*, 257–271.

 *materials*

Article

The Application of Functionalized Pillared Porous Phosphate Heterostructures for the Removal of Textile Dyes from Wastewater

José Jiménez-Jiménez [1], Manuel Algarra [1] , Vanessa Guimarães [2], Iuliu Bobos [2] and Enrique Rodríguez-Castellón [1,*]

[1] Departamento de Química Inorgánica, Facultad de Ciencias, Universidad de Málaga, Campus de Teatinos s/n, 29071 Málaga, Spain; jjimenez@uma.es (J.J.-J.); malgarra67@gmail.com (M.A.)
[2] Instituto de Ciências da Terra, Porto, DGAOT, Faculdade de Ciências, Universidade do Porto, Rua do Campo Alegre 687, 4169-007 Porto, Portugal; guimavs@gmail.com (V.G.); ibobos@fc.up.pt (I.B.)
* Correspondence: castellon@uma.es; Tel.: +34-952-131-873

Received: 26 July 2017; Accepted: 13 September 2017; Published: 21 September 2017

Abstract: A synthesized functionalized pillared porous phosphate heterostructure (PPH), surface functionalized phenyl group, has been used to remove the dye Acid Blue 113 from wastewater. X-ray photoemission spectroscopy XPS and X-ray diffraction (XRD) were used to study its structure. The specific surface area of this was 498 m^2/g. The adsorption capacities of PPH and phenyl surface functionalized (Φ PPH) were 0.0400 and 0.0967 mmol/g, respectively, with a dye concentration of 10^{-5} M when well fitted with SIPS and Langmuir isotherms respectively (pH 6.5, 25 °C). The incorporation of the dye to the adsorbent material was monitored by the S content of the dye. It is suggested as an alternative for Acid Blue 113 remediation.

Keywords: dye remediation; adsorption; azo dye; wastewater; pillared porous phosphate heterostructures; isotherm

1. Introduction

Nowadays, among the concerns that must be dealt with on a global level are toxic and carcinogenic environmental pollutants. A large portion of toxic contaminants are dyes, used for dyeing textiles and other industrial purposes. Many physical, chemical and biophysical processes have been studied for their ability to remove environmental pollutants [1]. In particular, the development of new technologies for the easy decolorization or remediation of these types of compounds is of current interest.

Dyes used in the textile industry include a wide range of chemical compounds, based mainly on substituted aromatic structures linked by azo groups [2]; therefore textile dye wastewater is based not only in color removal but also in the degradation, mineralization and total removal. A large variety of techniques has been used including chemical methods such as coagulation [3–5], flocculation [6–8] or precipitation [9,10], and new photocatalytic processes such as the Fenton reaction [11,12] are being implemented and optimized; biological with aerobic and anaerobic microbial degradation [13–15], and the use of pure enzymes [16,17] have been investigated to overcome physical methods such as membrane-filtration processes. Nano-filtration [18–20] reverse osmosis [21], electro-dialysis [22–24] and sorption techniques [25–30] are the most conventional methods due to higher production costs and regeneration difficulties. There is a trend towards developing sustainable water treatment materials and processes [31,32] and consequently, the possibilities of the use of new synthetic material—with low cost and easy availability—are desirable to be evaluated for wastewater management.

Functionalized mesoporous materials are interesting adsorbents in comparison with other commercial DVB or styrene base adsorbents because of their attractive properties, such as a high level of heavy metal removal capacity, large surface area, narrow pore size distribution and the specificity of the chelating agents inserted in the network. An emerging group, the synthesized materials are pillared clay structures (PILCs) and similar pillared porous phosphate heterostructures (PPH) [33,34]. These materials can be specifically designed for their use for Hg(II) and Ni(II) remediation from the electrochemical industry [35], or catalytic applications [36–40].

We investigate herein, for the first time to the best of our knowledge, the use of PPH modified with functional phenyl groups as a new class of adsorbents for dyes from wastewater. These solids are synthesized from different combinations of silicon precursors and surfactant along with various combinations of condensation processes of alkoxides. The effect of the adsorbate was studied in order to achieve optimal conditions for the adsorption of the azo dye Acid blue (AB113), proving to be an appropriate and cost effective alternative to remove it from wastewater.

2. Materials and Methods

2.1. Synthesis of Φ_x-PPH Adsorbent Materials

The synthesis of Φ_x-PPH materials was carried out as described elsewhere [35] by co-condensation of phenyl-triethoxysilane (PTES) with tetraethylorthosilicate (TEOS) in the interlayer region of zirconium phosphate. Thus, cetyltrimethylammonium (CTMA)-expanded zirconium phosphate (CTMAZrP) was prepared from a solution of CTMA-Br in 1-propanol, H_3PO_4 (85%) and zirconium(IV) propoxide (70%) according to previously reported procedures [33]. The CTMAZrP obtained was suspended in water (10 g/L) and a solution of hexadecylamine in 1-propano (0.145 M) was added as a co-surfactant. Increasing the surfactant content in the interlayer region improves the hydrolysis and condensation of the mixture of different alkoxides. After one day of stirring, a solution of TEOS (50%, v/v in 1-propanol) and the corresponding PTES—with TEOS/PTES molar ratios of 5, 25 and 50—were added, and in each case the resulting suspension was stirred at room temperature for three days. The obtained solids were centrifuged, washed with ethanol and dried at 60 °C in air. After this stage, silica galleries are formed between the layers of zirconium phosphate and the surfactant molecules are located inside these galleries. To release this space, it has proceeded to the removal of the surfactant by cation exchange with a solution of HCl in ethanol, because the calcination process is not a suitable procedure due to the removal of the phenyl functionalized silica walls in the galleries. After various tests, both the extractor solution concentration and the number of extractions were optimzed, yielding greater efficiency in the extraction process (three times) at a concentration of HCl:ethanol (1:10; $v:v$). Three materials with different TEOS/PTES molar ratio added ($\times$) are denoted as Φ_x-PPH. All reagents used were purchased from Sigma-Aldrich (Barcelona, Spain).

2.2. Adsorption Experiments AB113 in Solution Data Analysis

Batch adsorption experiments were performed at room temperature adding 50 mg of Φ_x-PPH at different volumes (a range of volumes was used) of 10^{-5} M of AB113. After one day, the solid was removed by centrifugation and the solution was analysed by UV-Vis spectroscopy. The Langmuir model assumes that the sorption sites are identical and energetically equivalent due to its homogeneous structure [41,42]. The equilibrium is obtained when the monolayer formation on the sorbent occurs. The Langmuir isotherm is described according to the following equation (Equation (1)):

$$q_e = \frac{q_m K_L C_e}{1 + K_L C_e} \tag{1}$$

where, q_e (mol/kg) and C_e (mol/L) are the equilibrium concentrations of AB113 in the solid and the liquid phase, respectively, q_m (mmol/g) is the maximum sorption capacity and K_L (L/kg) is the Langmuir constant related to the energy of adsorption. q_e is obtained according to Equation (2):

$$q_e = \left(C_i - C_f\right)\frac{V}{m} \tag{2}$$

where, C_i and C_f are the concentrations of AB113 at the beginning and the end of the adsorption process, V is the solution volume used during batch experiments (a range of volumes was used) and m is the mass of Φ_5-PPH used.

The Freundlich model was used to describe the adsorption of contaminants on heterogeneous surfaces consisting of sites with different exponential distribution and energies [41,42]. The equation of the Freundlich sorption isotherm is expressed by Equation (3):

$$q_e = K_F C_e^n \tag{3}$$

where, K_F and n are the Freundlich adsorption isotherm constants, being indicative of the adsorption extension and the degree of the surface heterogeneity. The SIPS isotherm combines both Freundlich and Langmuir isotherms, which at low adsorbate concentration behaves as the Freundlich isotherm [41] and at high concentration it predicts a monolayer adsorption capacity characteristic to the Langmuir model [41]. The mathematical representation of this model is given by Equation (4):

$$q_e = \frac{q_m (K_s C_e)^n}{1 + (K_s C_e)^n} \tag{4}$$

where, q_m (mol/kg) is the maximum adsorption capacity, which can also be expressed as Nt, a measure of the total number of binding sites available per g of sorbent, K_S is the affinity constant for adsorption (L/kg) and n is the Freundlich parameter that takes into account the system heterogeneity. The SIPS isotherm is reduced to the Langmuir form for $n = 1$ and a homogeneous surface is considered. The greater the difference from this value, the greater the solid surface heterogeneity.

2.3. Characterization Methods

UV-Vis measurements were carried out with a Shimadzu UV-1800 spectrophotometer (Shimadzu Corporation, Kyoto, Japan). Powder diffraction patterns were collected on an X'Pert Pro MPD automated diffractometer (PANalytical, Almelo, The Netherlands) equipped with a Ge(111) primary monochromator (strictly monochromatic Cu-K$_\alpha$ radiation) and an X'Celerator detector (PANalytical, Almelo, The Netherlands). Textural parameters were obtained from N_2 adsorption-desorption isotherms with a Micromeritics ASAP 2020 (Micromeritics Ltd., Bedfordshire, UK). The specific surface area and pore volume of the Φ_x-PPH were determined by the adsorption-desorption of N_2 at -196 °C. Before analysis, the Φ_x-PPH samples were degassed at 150 °C up to 10^{-4} Torr. Pore volume and an average pore diameter were determined by the Barrett, Joyner, Halenda model. X-ray photoelectron spectra (XPS) were obtained using a Physical Electronics PHI 5700 spectrometer with a non-monochromatic Mg K$_\alpha$ radiation (300 W, 15 kV, hv = 1256.3 eV) as the excitation source. Spectra were recorded at 45° take-off angles by a concentric hemispherical analyser (Physical Electronics, Inc., Chanhassen, MN, USA) operating in the constant pass energy mode at 25.9 eV, using a 720 µm diameter analysis area. Under these conditions the Au $4f_{7/2}$ line was recorded with 1.16 eV FWHM at a binding energy of 84.0 eV. The spectrometer energy scale was calibrated using Cu $2p_{3/2}$, Ag $3d_{5/2}$ and Au $4f_{7/2}$ photoelectron lines at 932.7, 368.3 and 84.0 eV, respectively. Charge referencing was done against adventitious carbon (C $1s$ 284.8 eV). The C $1s$ signal of some samples was also studied with an Al K$_\alpha$ radiation due to the presence of sodium to avoid the overlapping of the Na KLL Auger signal at 290.3 eV. Solid surfaces were mounted on a sample holder without adhesive tape and kept overnight in a high vacuum in the preparation chamber before they were transferred to the analysis chamber of the spectrometer. Each region was scanned with several sweeps until a good signal-to-noise ratio was observed. The pressure in the analysis chamber was maintained lower than 10^{-9} Torr. A Shirley-type background was subtracted from the signals. Recorded spectra were always fitted using Gauss-Lorentz curves in order to determinate more accurately the binding energy of the different element core levels. The accuracy of binding energy (BE) values was within ± 0.1 eV.

3. Results and Discussion

3.1. Adsorbent Characterization

Elemental analysis does not provide enough information to determine the presence of phenyl groups. The residual surfactant and the alcohol adsorbed used in the extraction procedure can interfere in the carbon content (%C). Therefore, the phenyl incorporation cannot be determined by CHN elemental analysis. However, the increasing %C when phenyl groups are grafted was observed. Thus, for the extracted materials in which the amount of surfactant present is residual, the observed differences in the %C are remarkable, being this C % higher for material Φ_5-PPH (9.04%) and reduced for Φ_{50}-PPH (2.62%), as displayed in Table 1.

Table 1. Chemical composition of Φ-PPH material.

Material	%C	%N	%C *	%N *
Φ_5-PPH	9.04	0.147	19.53	1.30
Φ_{25}-PPH	3.81	0.217	18.91	1.27
Φ_{50}-PPH	2.62	0.222	18.53	1.28

* Previous surfactant extraction.

The concentration of phenyl groups in the synthesized material was determined by UV-VIS spectroscopy at λ = 210 nm in hexane. As seen in the UV-VIS spectra (Figure 1), the absorption due to the phenyl groups decreases with the TEOS/Φ molar ratio. A calibration curve was performed by hydrolysis of different amounts of phenyl triethoxysilane and the amount of phenyl groups incorporated in each material was determined (Table 2). As expected, the highest observed incorporation of phenyl groups occurred with sample Φ_5-PPH, although the found value (0.047 mmol/g) was below the theoretical one if all the phenyl groups added were incorporated to the structure. This fact may be due to steric effects and to the non-polar character of phenyl groups which prevents their presence in large amounts in the interlayer space, giving a low incorporation when the silica galleries are formed upon hydrolysis and condensation of PTES, together with the TEOS.

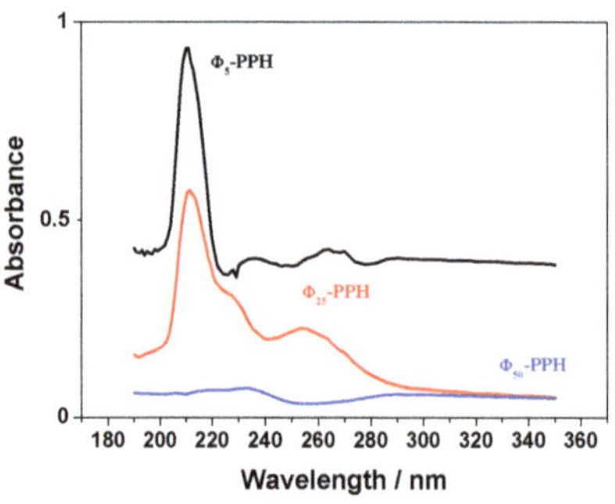

Figure 1. UV-Vis spectra of Φ_5-PPH, Φ_{25}-PPH and Φ_{50}-PPH at 210 nm in hexane.

Table 2. Characteristics of the different materials synthesized.

Material	mmol Φ/g	mmol P/g	Φ/P (Exp)	Φ/P (Exp)	S_{BET} (m²/g)	V_p (cm³/g)	C_{BET}	S_{ac} (m²/g) *	ΣV_p (cm³/g) •	Dp (Å)
Φ_5 –PPH	0.047	1.73	0.027	0.6	498	0.516	133	482	0.472	39.13
Φ_{25}– PPH	0.042	1.86	0.023	0.12	545	0.596	123	540	0.565	41.87
Φ_{50}– PPH	0.003	1.78	0.002	0.06	571	0.657	133	546	0.609	44.59

* Previous tensioactive extraction and • Determinated by the Cranston e Inkley method.

This also justifies the found number of functional groups incorporated into the material Φ_{25}-PPH that is only slightly lower than that found for Φ_5-PPH, despite the fact that the amount of functionalized alkoxide added to the latter was five times the added amount in the case of Φ_{25}-PPH. However, in spite of the increase in the number of phenyl groups incorporated, the final absorbance remained constant. Finally, as expected, in the case of material Φ_{50}-PPH, the incorporation of phenyl groups in this material is very low.

After removing the surfactant used as a template, the inner silica galleries are free and a porous material is obtained. For hybrid Φ_x-PPH, the surface phenyl groups are directed to the inner porous material. This is true if the inorganic framework is preserved after surfactant removal.

This fact can be determined by studying the corresponding XRD diffractograms. The results show a single broad signal at a low angle corresponding to the d_{001} reflection, showing the separation of the layers of zirconium phosphate (Figure 2A). This confirms the presence of silica galleries in the interlayer region of zirconium phosphate, which keep these layers separate when the surfactant has been removed.

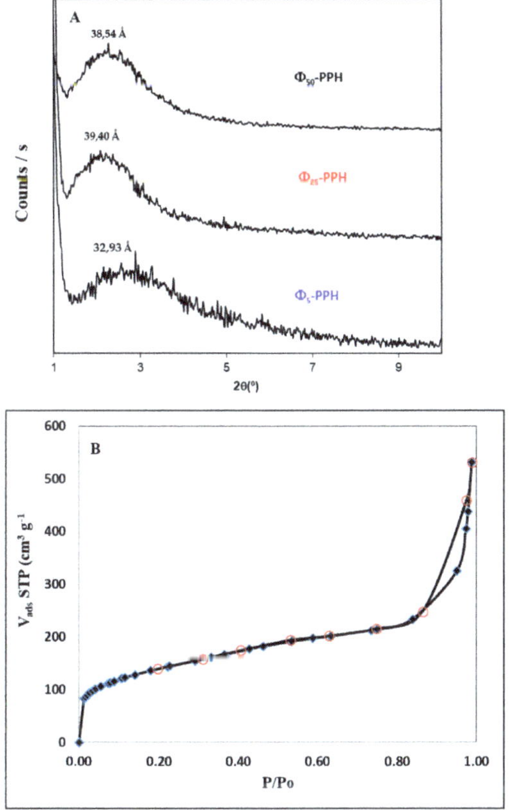

Figure 2. (A) XRD pattern of Φ_5-PPH, Φ_{25}-PPH and Φ_{50} PPH; (B) N_2 adsorption-desorption isotherms at 77 K of Φ_5-PPH.

In Φ_5-PPH, the diffraction peak appears at a slightly lower angle so that the basal spacing is 32.9 Å, the intensity is weak and the definition of the peak poor. This indicates that the presence of phenyl groups hinders the arrangement of the internal phase material. This low crystallinity can be justified by the steric hindrance of the phenyl groups, due to their non-polar character, since the surfactant micelles which act as structural elements of the silica galleries present a polar end placed in this environment which hinders the correct formation of the inorganic structure, mainly for the precursors which contain the phenyl group. This fact is reflected, as discussed in the previous section, in the incorporation of organic groups where the amount of phenyl groups is lower than the other functionalized PPH studied [35].

The evaluation of the typical textural parameters such as surface area and pore size distribution will confirm the formation of silica in the galleries of the interlayer region. Textural parameters obtained from the corresponding Nitrogen adsorption-desorption isotherms of N_2 at $-196\ ^\circ$C correspond to type IV (IUPAC) [36,37], characteristic of mesoporous materials (Figure 2B, Table 2). The surface area values (S_{BET}) obtained are quite high (above 500 m^2/g), indicating a large accessible surface within the material due to the presence of silica galleries in the interlayer region being lower than the pure heterostructures (620 m^2/g) [33], showing that the incorporation of phenyl groups decreases the surface area of the material. Regarding the volume of pores (V_p), these materials have very high values, even more than the pure silica heterostructures (0.543 cm^3/g). This may be due to the presence of larger pore diameters as shown in the distribution of pores.

The surface chemical composition of the adsorbent Φ_5-PPH before and after contact with AB113 (sample Φ_5-PPH + AB113) is shown in Table 3. Upon contact with the dye, the percentages of C and N increase and S are now detected. However, Na was not detected indicating that the dye anion was taken up. The empiric formula of the dye anion is $C_{32}H_{21}N_5O_6S_2$ and the theoretical N/S atomic ratio is 2.5. The observed N/S atomic ratio after incorporation of the dye is 2.6, a value very near the theoretical one indicating the incorporation of the dye. Sulfur is the only element of the dye that is not present in the adsorbent Φ_5-PPH and the S 2p core level spectrum for sample Φ_5-PPH + AB113 shows a S 2$p_{3/2}$ contribution at 168.2 eV assigned to the presence of S(VI) as the sulfonic group of the dye [43].

Table 3. Atomic concentration (%) of samples Φ_5-PPH and Φ_5-PPH + AB113 determined by XPS.

Sample	C	N	O	Si	P	Zr	S
Φ_5-PPH	24.50	0.50	50.07	19.55	3.52	1.86	-
Φ_5-PPH + AB113	29.56	0.94	46.00	18.42	2.19	1.69	0.36

3.2. Adsorption Isotherms

From the results obtained after chemical, structural and textural characterization, Φ_5-PPH material was selected as adsorbent for AB113, given that it presents the highest phenyl group surface density (Table 1). Also, to evaluate the effect of the phenyl group on the AB113 adsorption, the pristine PPH material was used as a reference. AB113 adsorption isotherms were investigated in triplicate at 25 $^\circ$C and pH 6.5.

Figure 3 shows the fitting of the experimental data to the adsorption isotherms for PPH and Φ_5-PPH respectively. In the case of pristine PPH, the isotherm was fitted to the SIPS isotherm with a saturated adsorption capacity of 0.04 mmol/g (Figure 3A). The adsorption data from Φ_5-PPH were fitted to a Langmuir isotherm model, showing a saturated adsorption capacity (Q_o) of 0.0967 mmol/g and a Langmuir constant (K_L) of 613 L/Kg (Figure 3B). Other model fittings are shown as supplementary information as Figure S1.

The adsorption effectiveness was also evaluated. The retention effectiveness was close to 100% for Φ_5-PPH material, when the adsorption capacity was lower than 0.040 mmol/g. From this point it decreases while increasing the amount of AB113 in the solution. For the pure PPH material, the highest retention percentage reached was 55% for the first fit, decreasing in the next points. In the literature, AB113 adsorption on mesoporous activated carbon is reported [44], where the observed adsorption capacities are in the order obtained for Φ_5-PPH. In this case the adsorption capacities for an activated carbon obtained from a rubber tyre and for a commercial activated carbon obtained from the Langmuir isotherm (Q_o), were 0.014 and 0.011 mmol/g, respectively, although for both activated carbons the specific surface areas are higher than that observed for Φ_5-PPH with values of 562 and 1168 m^2/g. Another adsorbent based in red mud obtained a Q_o = 0.112 mmol/g [45], demonstrating the feasibility of the proposed system to be applied.

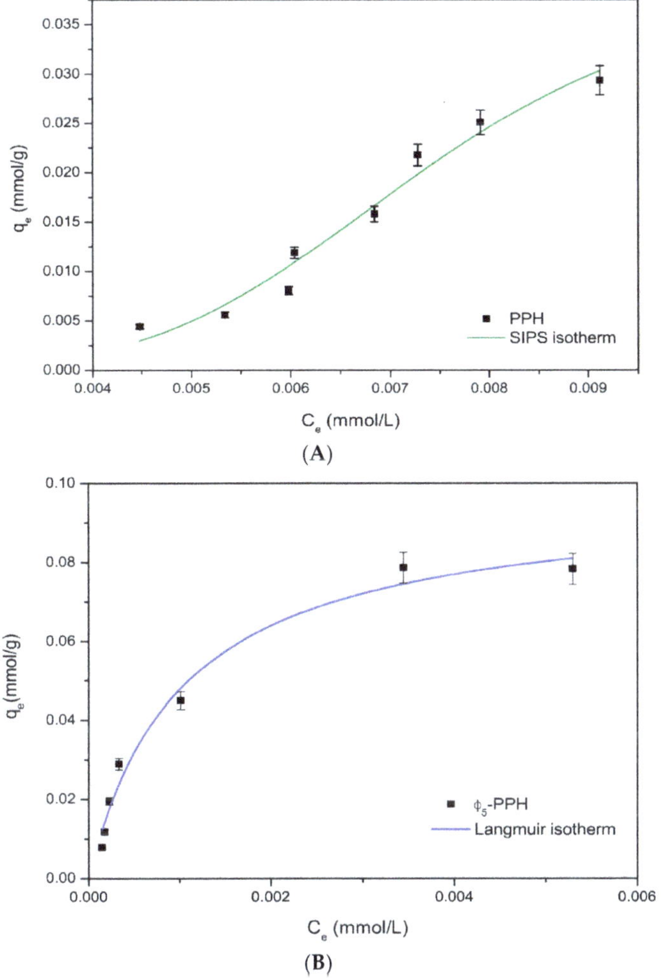

Figure 3. AB113 adsorption isotherms in (**A**) PPH and (**B**) Φ_5-PPH and in relation to mmol AB113 for g of sorbent.

The difference in the adsorption capacities of PPH and Φ_5-PPH can be explained by the difference in the intermolecular forces between the different surface groups and the dye monomer. In the case of PPH, the main forces involved are strong hydrogen bonding with the AB113 (Si-OH and P-OH) and the presence of aromatic rings on the surface of phenyl-functionalized Φ_5-PPH which offer a superior degree of delocalization due to the π-π stacking of the phenyl surface groups of the adsorbent (Figure 4). The different mechanisms of adsorption are also indicated by the fit of sorption data to two distinct isothermal models. The better fit of the Φ_5-PPH sorption data to the Langmuir isotherm reveals that the sorption sites are identical and energetically homogeneous, suggesting that the coordination between AB113 and the aromatic rings of Φ_5-PPH is the main mechanism of adsorption involved.

However, the better fit of PPH sorption data to the SIPS isotherm indicates that there is more than one type of site involved in this process, each characterized by distinct energies and distinct affinities to adsorb. In this case, it is particularly related to the adsorption on the Si-OH and P-OH groups of PPH. To demonstrate the mechanism described, decreasing a specific area in our material does not limit increased dye adsorption.

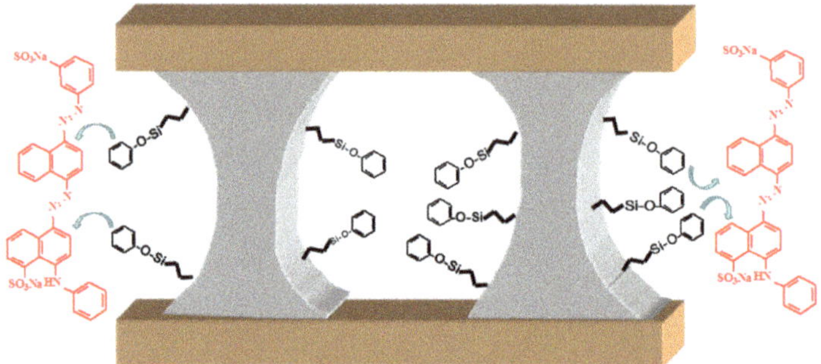

Figure 4. Proposed mechanism of interaction between Φ_5-PPH and AB113.

The various applications of phenyl functionalized adsorbent have been reported in recent literature [46–49]. However, the current work is unique in terms of the use of phenyl-PPH for dye adsorption, which is reported here for the first time.

3.3. Application in Wastewater

In accordance with the results previously obtained, Φ_5-PPH was used as an adsorbent material for treating real wastewater from a local textile stamp factory. Figure 5 shows the absorbance at 270 nm of water from a rinse step treated with different mass sorbent, where the absorbance at 270 nm achieves the minimum from a mass/volume ratio of 10 g/m^3, which indicates that this type of hybrid phenyl material can be a promising dye scavenger.

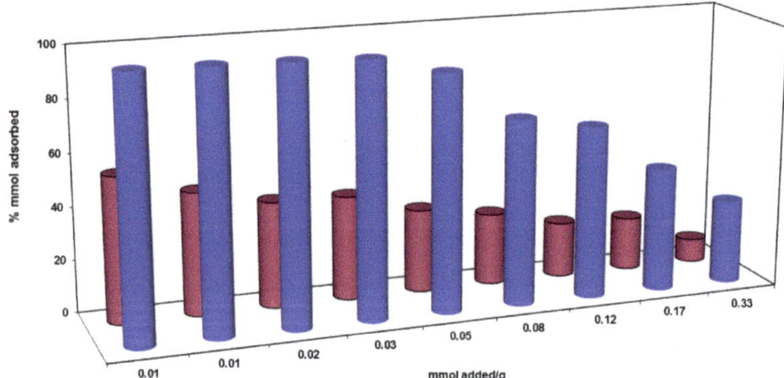

Figure 5. Remediation of an azo dye from industrial textile wastewater by Φ_5-PPH recorded at 270 nm (■) and compared with raw PPH (■).

4. Conclusions

The functionalized porous phosphate heterostructure materials with phenyl functionalized surface proved to be effective adsorbents for the removal of AB113 azo dye from wastewater from the textile industry. With a surface area (SBET) of 500 m^2/g, the best known adsorption process is Langmuir, compared to the other tested models such as Freundlich and SIPS, showing saturation (Q_o) of 0.0967 mmol/g. An adsorption mechanism is proposed between the phenyl groups of the Φ_5-PPH and AB113, based on the affinity between both groups due to π-π stacking, suggested by the Langmuir model. The results indicate that Φ_5-PPH is a good alternative to remove the azo dye from industrial wastewater, as shown in the application studied and supported by XPS analysis of the surface of Φ_5-PPH after the application of the N/S ration.

Supplementary Materials: The following are available online at www.mdpi.com/1996-1944/10/10/1111/s1, Figure S1: Fitting of experimental adsorption of AB113 on Φ_5-PPH.

Acknowledgments: This work was supported by the project MINECO CTQ2015-68951-C3-3-R and FEDER funds. V. Guimarães thanks the PhD scholarship (Ref. SFRH/BD/79969/2011) financed by Fundação para a Ciência e Tecnologia (FCT), Portugal.

Author Contributions: Authors have contributed equally in all the work.

Conflicts of Interest: The authors declare no conflict of interest.

References

1. Lichtfouse, E.; Schwarzbauer, J.; Robert, D. *Environmental Chemistry. Green Chemistry and Pollutants in Ecosystems*; Springer: Berlin, Germany, 2005.

2. Hunger, K. (Ed.) *Industrial Dyes. Chemistry, Properties, Applications*; Wiley-VCH: Weinheim, Germany, 2003.

3. Shi, B.; Li, G.; Wang, D.; Feng, C.; Tang, H. Removal of direct dyes by coagulation: The performance of preformed polymeric aluminum species. *J. Hazard. Mater.* **2007**, *143*, 567–574. [CrossRef] [PubMed]

4. Alinsafi, A.; Khemis, M.; Pons, M.N.; Leclerc, J.P.; Yaacoubi, A.; Benhammou, A.; Nejmeddine, A. Electro-coagulation of reactive textile dyes and textile wastewater. *Chem. Eng. Process. Process Intensif.* **2005**, *44*, 461–470. [CrossRef]

5. Cañizares, P.; Martínez, F.; Jiménez, C.; Lobato, J.; Rodrígo, M.A. Coagulation and electrocoagulation of wastes polluted with dyes. *Environ. Sci. Technol.* **2006**, *40*, 6418–6424. [CrossRef] [PubMed]

6. Zonoozi, M.H.; Moghaddam, M.R.; Arami, M. Coagulation/flocculation of dye-containing solutions using polyaluminium chloride and alum. *Water Sci. Technol.* **2009**, *59*, 1343–1351. [CrossRef] [PubMed]

7. Moghaddam, S.S.; Moghaddam, M.R.A.; Arami, M. Coagulation/flocculation process for dye removal using sludge from water treatment plant: Optimization through response surface methodology. *J. Hazard. Mater.* **2010**, *175*, 651–657. [CrossRef] [PubMed]

8. Petzold, G.; Schwarz, S.; Mende, M.; Jaeger, W. Dye flocculation using polyampholytes and polyelectrolyte-surfactant nanoparticles. *J. Appl. Polym. Sci.* **2007**, *104*, 1342–1349. [CrossRef]

9. Lemlikchi, W.; Sharrock, P.; Mecherri, M.O.; Fiallo, M.; Nzihou, A. Reaction of calcium phosphate with textile dyes for purification of wastewaters. *Desalin. Water Treat.* **2014**, *52*, 1669–1673. [CrossRef]

10. Raïs, Z.; El Hassani, L.; Maghnouje, J.; Hadji, M.; Ibnelkhayat, R.; Nejjar, R.; Kherbeche, A.; Chaqroune, A. Dyes' removal from textile wastewater by phosphogypsum using coagulation and precipitation method. *Phys. Chem. News* **2002**, *7*, 100–109.

11. Lucas, M.S.; Algarra, M.; Jiménez-Jiménez, J.; Rodríguez-Castellón, E.; Péres, J.A. Catalytic Activity of Porous Phosphate Heterostructures-Fe towards Reactive Black 5 Degradation. *Int. J. Photoenergy* **2013**, *2013*, 658231. [CrossRef]

12. Lucas, M.S.; Tavares, P.B.; Péres, J.A.; Farias, J.L.; Rocha, M.; Pereira, C.; Freire, C. Photocatalytic degradation of Reactive Black 5 with TiO$_2$-coated magnetic nanoparticles. *Catal. Today* **2013**, *209*, 116–121. [CrossRef]

13. Senan, R.C.; Shaffiqu, T.S.; Roy, J.J.; Abaham, T.E. Aerobic Degradation of a Mixture of Azo Dyes in a Packed Bed Reactor Having Bacteria Coated Laterite Pebbles. *Biotechnol. Prog.* **2003**, *19*, 647–651. [CrossRef] [PubMed]

14. De los Cobos Vasconcelos, D.; Ruiz-Ordaz, N.; Galíndez-Mayer, J.; Galíndez-Mayer, H.; Juàrez-Ramírez, C.; Aarón, L.M. Aerobic biodegradation of a mixture of sulfonated azo dyes by a bacterial consortium immobilized in a two-stage sparged packed-bed biofilm reactor. *Eng. Life Sci.* **2012**, *12*, 39–48. [CrossRef]

15. O'Neill, C.; López, A.; Esteves, S.; Hawkes, F.R.; Hawkes, D.L.; Wilcox, S. Azo-dye degradation in an anaerobic-aerobic treatment system operating on simulated textile effluent. *Appl. Microbiol. Biotechnol.* **2000**, *53*, 249–254. [CrossRef] [PubMed]

16. Shah, P.D.; Dave, S.R.; Rao, M.S. Enzymatic degradation of textile dye Reactive Orange 13 by newly isolated bacterial strain Alcaligenes faecalis PMS-1. *Int. Biodeterior. Biodegrad.* **2012**, *69*, 41–50. [CrossRef]

17. Shaffiqu, T.S.; Roy, J.J.; Nair, R.A.; Abraham, T.E. Degradation of textile dyes mediated by plant peroxidases. *Appl. Biocem. Biotechnol.* **2002**, *102–103*, 315–326. [CrossRef]

18. Koyuncu, I. Reactive dye removal in dye/salt mixtures by nanofiltration membranes containing vinylsulphone dyes: Effects of feed concentration and cross flow velocity. *Desalinisation* **2002**, *143*, 243–253. [CrossRef]

19. Dul'neva, T.Y.; Titoruk, G.N.; Kucheruk, D.D.; Goncharuk, V.V. Purification of wastewaters of direct dyes by ultra-and nanofiltration ceramic membranes. *J. Water Chem. Technol.* **2013**, *35*, 165–169. [CrossRef]

20. Gomesa, A.C.; Gonçalves, I.C.; de Pinho, M.N. The role of adsorption on nanofiltration of azo dyes. *J. Membr. Sci.* **2005**, *255*, 157–165. [CrossRef]

21. Uzal, N.; Yilmaz, L.; Yetis, U. Nanofiltration and reverse osmosis for reuse of indigo dye rinsing waters. *Sep. Sci. Technol.* **2010**, *45*, 331–338. [CrossRef]

22. Körbahti, B.K.; Artut, K.; Geçgel, C.; Özer, A. Electrochemical decolorization of textile dyes and removal of metal ions from textile dye and metal ion binary mixtures. *Chem. Eng. J.* **2011**, *173*, 677–688. [CrossRef]

23. Cheng, S.; Oatley, D.L.; Williams, P.M.; Wright, C.J. Characterisation and application of a novel positively charged nanofiltration membrane for the treatment of textile industry wastewaters. *Water Res.* **2012**, *46*, 33–42. [CrossRef] [PubMed]

24. Chandramowleeswaran, M.; Palanivelu, K. Treatability studies on textile effluent for total dissolved solids reduction using electrodialysis. *Desalination* **2006**, *201*, 164–174. [CrossRef]

25. Ho, Y.S.; McKay, G. Sorption of dye from aqueous solution by peat. *Chem. Eng. J.* **1998**, *70*, 115–124. [CrossRef]

26. Delval, F.; Crini, G.; Morin, N.; Vebrel, J.; Bertini, S.; Torri, G. The sorption of several types of dye on crosslinked polysaccharides derivatives. *Dyes Pigm.* **2002**, *53*, 79–92. [CrossRef]

27. Zhu, M.X.; Li, Y.P.; Xie, M.; Xin, H.Z. Sorption of an anionic dye by uncalcined and calcined layered double hydroxides: A case study. *J. Hazard. Mater.* **2005**, *120*, 163–171. [CrossRef] [PubMed]

28. Mon, J.; Flury, M.; Harsh, J.B. Sorption of four triarylmethane dyes in a sandy soil determined by batch and column experiments. *Geoderma* **2006**, *133*, 217–224. [CrossRef]

29. Wawrzkiewicz, M. Removal of C.I. Basic Blue 3 dye by sorption onto cation exchange resin, functionalized and non-functionalized polymeric sorbents from aqueous solutions and wastewaters. *Chem. Eng. J.* **2013**, *217*, 414–425. [CrossRef]

30. Široký, J.; Blackburn, R.S.; Bechtold, T.; Taylor, J.; White, P. Alkali treatment of cellulose II fibres and effect on dye sorption. *Carbohydr. Polym.* **2011**, *84*, 299–307. [CrossRef]

31. Razali, M.; Kim, J.F.; Attfield, M.; Budd, P.M.; Drioli, E.; Lee, Y.M.; Szekely, G. Sustainable wastewater treatment and recycling in membrane manufacturing. *Green Chem.* **2015**, *17*, 5196. [CrossRef]

32. Qu, X.; Brame, J.; Li, Q.; Alvarez, P.J.J. Nanotechnology for a safe and sustainable water supply: Enabling integrated water treatment and reuse. *Acc. Chem. Res.* **2013**, *46*, 834–843. [CrossRef] [PubMed]

33. Jiménez-Jiménez, J.; Rubio-Alonso, M.; Eliche-Quesada, D.; Rodriguez-Castellón, E.; Jiménez-López, A. Synthesis and characterisation of acid mesoporous phosphate heterostructure (PPH) materials. *J. Mater. Chem.* **2005**, *15*, 3466. [CrossRef]

34. Jiménez-Jiménez, J.; Rubio-Alonso, M.; Eliche-Quesada, D.; Castellón, E.R.; Jiménez-López, A. Synthesis and characterization of mixed silica/zirconia and silica/titania porous phospate heterostuctures (PPH). *J. Phys. Chem. Solids* **2006**, *67*, 1007–1010. [CrossRef]

35. Jiménez-Jiménez, J.; Algarra, M.; Castellón, E.R.; Jiménez-López, A.; Esteves da Silva, J.C.G. Hybrid porous phosphate heterostructures as adsorbents of Hg(II) and Ni(II) from industrial sewage. *J. Hazard. Mater.* **2011**, *190*, 694–699. [CrossRef] [PubMed]

36. Pierotti, R.A.; Rouquerol, J. Reporting Physisorption Data for Gas/Solid Systems with Special Reference to the Determination of Surface Area and Porosity. *Pure Appl. Chem.* **1985**, *57*, 603.

37. Lowell, S. *Characterization of Porous Solids and Powders: Surface Area, Pore Size and Density*; Springer: Berlin, The Netherlands, 2004.

38. Moreno-Tost, R.; Oliveira, M.L.; Eliche-Quesada, D.L.; Jiménez-Jiménez, J.; Jiménez-López, A.; Rodríguez-Castellón, E. Evaluation of Cu-PPHs as active catalysts for the SCR process to control NOx emissions from heavy duty diesel vehicles. *Chemosphere* **2008**, *72*, 608–615. [CrossRef] [PubMed]

39. Eliche-Quesada, D.; Macias-Ortiz, M.I.; Jiménez-Jiménez, J.; Rodríguez-Castellón, E.; Jiménez-López, A. Catalysts based on Ru/mesoporous phosphate heterostructures (PPH) for hydrotreating of aromatic hydrocarbons. *J. Mol. Catal. A* **2006**, *255*, 41–48. [CrossRef]

40. Soriano, M.D.; Jiménez-Jiménez, J.; Concepción, P.; Jiménez-López, A.; Rodríguez-Castellón, E.; López Nieto, J.M. Selective oxidation of H2S to sulfur over vanadia supported on mesoporous zirconium phosphate heterostructure. *Appl. Catal. B Environ.* **2009**, *92*, 271–279. [CrossRef]

41. Freundlich, H.; Heller, W. On Adsorption in Solution. *J. Am. Chem. Soc.* **1939**, *61*, 2228. [CrossRef]

42. Sips, R. On the structure of a catalyst surface. *J. Chem. Phys.* **1948**, *16*, 490. [CrossRef]

43. Seredych, M.; Rodríguez-Castellón, E.; Biggs, M.J.; Skinner, W.; Bandosz, T.J. Effect of visible light and electrode wetting on the capacitive performance of S- and N-doped nanoporous carbons: Importance of surface chemistry. *Carbon* **2014**, *78*, 540–558. [CrossRef]

44. Gupta, V.K.; Gupta, B.; Rastogi, A.; Agarwal, S.; Nayak, A. A comparative investigation on adsorption performances of mesoporous activated carbon prepared from waste rubber tire and activated carbon for a hazardous azo dye—Acid Blue 113. *J. Hazard. Mater.* **2011**, *186*, 891–901. [CrossRef] [PubMed]

45. Shirzad-Siboni, M.; Jafari, S.J.; Giahi, O.; Kim, I.; Lee, S.M.; Yang, J.K. Removal of acid blue 113 and reactive black 5 dye from aqueous solutions by activated red mud. *J. Ind. Eng. Chem.* **2014**, *20*, 1432–1437. [CrossRef]

46. Schumacher, C.; Seaton, N.A. Modeling of Organically Functionalized Mesoporous Silicas for the Design of Adsorbents. *Adsorption* **2005**, *11*, 643–648. [CrossRef]

47. Huang, D.; Sha, Y.; Zheng, S.; Liu, B.; Deng, C. Preparation of phenyl group-functionalized magnetic mesoporous silica microspheres for fast extraction and analysis of acetaldehyde in mainstream cigarette smoke by gas chromatography–mass spectrometry. *Talanta* **2013**, *115*, 427–434. [CrossRef] [PubMed]

48. Saad, N.; Al-Mawla, M.; Moubarak, E.; Al-Ghoul, M.; El-Rassy, H. Surface-functionalized silica aerogels and alcogels for methylene blue adsorption. *RSC Adv.* **2015**, *5*, 6111. [CrossRef]

49. Saraji, M.; Khaje, N. Phenyl-functionalized silica-coated magnetic nanoparticles for the extraction of chlorobenzenes, and their determination by GC-electron capture detection. *J. Sep. Sci.* **2013**, *36*, 1090–1096. [CrossRef] [PubMed]

 *materials*

Article

Modification of 13X Molecular Sieve by Chitosan for Adsorptive Removal of Cadmium from Simulated Wastewater

Yan Shi [1], Ken Sun [1], Lixin Huo [1], Xiuxiu Li [2], Xuebin Qi [3,*] and Zhaohui Li [4,*]

[1] School of Environmental and Municipal Engineering, North China University of Water Resources and Electric Power, Zhengzhou 450045, China; shiyan@ncwu.edu.cn (Y.S.); sunken@ncwu.edu.cn (K.S.); 18748975624@163.com (L.H.)

[2] Henan Province key Laboratory of Water-Saving Agriculture, North China University of Water Resources and Electric Power, Zhengzhou 450045, China; lixiuxiu@163.com

[3] Farmland Irrigation Research Institute, CAAS, Xinxiang 453002, China

[4] Geosciences Department, University of Wisconsin—Parkside, Kenosha, WI 53144, USA

* Correspondence: qxb6301@yahoo.com.cn (X.Q.); li@uwp.edu (Z.L.)

Received: 9 August 2017; Accepted: 13 September 2017; Published: 19 September 2017

Abstract: Chitosan was used to modify a 13X molecular sieve to improve its cadmium removal capability. After being modified with 2% chitosan-acetate for 2 h at 30 °C, significant uptake of Cd^{2+} could be achieved. The uptake of Cd^{2+} on the modified 13X molecular sieve followed the Langmuir isotherms with a capacity of 1 mg/g. The kinetics of Cd^{2+} removal by modified 13X molecular sieve followed a pseudo second-order reaction, suggesting chemisorption or surface complexation. The Cd^{2+} removal with a sorbent dose of 2 g/L from an initial concentration of 100 µg/L reached more than 95% in 90 min. The equilibrium Cd^{2+} concentration was <5 µg/L, which meets the requirements of "Standards for Irrigation Water Quality" (GB5084-2005) (10 µg/L) and MCL and MCLG for groundwater and drinking water (5 µg/L) set by United States Environmental Protection Agency.

Keywords: cadmium; chitosan; modification; 13X molecular sieve; removal

---

1. Introduction

Cadmium (Cd) is a major water pollutant. With a high toxicity, Cd can cause chronic intoxication with an incubation period from 10 to 30 years. The half-life of Cd in the human body also varies between 10 and 30 years. People can develop gastroenteropathy or even hepatopathy, lung cancer, kidney disease, with continuing intake of food or water contaminated with Cd [1–4]. Currently, the USEPA has set both the maximum contamination level (MCL) and maximum contamination level goal (MCLG) at 5 µg/L [5]. Although its concentration in unpolluted natural waters was usually below 1 µg/L [6], a maximum value of 100 µg/L was reported in Rio Rimao in Peru [7]. In addition, the presence of Cd as an impurity in zinc in galvanized pipes may have potential for drinking water contamination. A mean concentration of 1–26 µg/L was found in samples of potable water in Saudi Arabia [8].

Cd cannot be microbiologically degraded in the natural environment, and can only be dispersed, enriched, or converted between its various forms [9]. The migration process of Cd in water is mainly dependent on precipitation, complexation, and adsorption. Although natural zeolite has high cation exchange capacities (CEC), its internal channels of different sizes make the uptake of heavy metal ions more kinetically controlled [10,11]. In addition, natural zeolite may have other impurities that can make water purification less effective. Compared with natural zeolite, synthetic zeolite has simpler phases and compositions, with less impurity, which avoids secondary pollution. Its internal pore channels and holes would be of uniform size, which can be regulated by adjusting the synthesis conditions, as well as larger internal and external surface area, good physical and chemical stability, exchange, adsorption, and catalysis capacities make it a good candidate for water treatment [12,13]. In particular, the synthetic 13X zeolite molecular sieve (MS) has a cubic crystal system with 3D pore channels, which can provide a faster intracrystalline diffusion for adsorption and catalysis [14,15]. However, its uptake of cationic heavy metals may be low.

Containing a large number of amino and hydroxyl groups, chitosan has high hydrophilicity and can absorb heavy metals in wastewater [16–18]. Chitosan-modified natural zeolite can achieve good removal performance of heavy metals (mostly Cu(II), Fe(III), Mn(II), Zn(II)) from wastewater with high concentrations [19–26]. However, little research has been conducted into the treatment of wastewater with low Cd^{2+} concentrations using chitosan-modified 13X molecular sieves (CMS). Therefore, in this study, conditions were optimized for the modification of MS by chitosan, and the CMS was tested for the removal of low concentrations of Cd^{2+} from simulated wastewater, so that the Cd^{2+} concentration would meet the discharge standards of 10 µg/L set by the Standards for Irrigation Water Quality (GB5084-2005), and 5 µg/L of MCL and MCLG set by USEPA. The results also provide technical support for safe use of unconventional water in agriculture.

2. Materials and Methods

2.1. Materials

The 13X molecular sieve was off-white in color with a grain size of 2 to 3 mm, and was purchased from Zhengyuan Haoye Chemical Technology Co. (Tianjing, China). Chitosan (food grade with 95% deacetylation) was purchased from Zhengzhou Mingrui Food Ingredients, Co., Ltd. (Zhengzhou, China) Nitric acid, hydrochloric acid, sodium hydroxide, glacial acetic acid, and Cd^{2+} stock solution (1.000 g/L) were all of reagent grade or analytical pure and were purchased from Zhiyuan Chemical Technology Co. (Tianjing, China).

2.2. Preparation of CMS

The MS was washed with tap water several times to remove the impurities, and then washed with deionized water 2–3 times before being dried at 105 °C for modification. The MS was added to chitosan solutions of different concentrations (0.5–3.0%) under acidic conditions to increase the solubility of chitosan [16,18]. The mixture was shaken under a constant temperature of 30 °C at 120 rpm for 30, 60, 90, 120, 150, and 180 min, before being washed with deionized water to neutral condition and dried at 55 °C.

2.3. Cd^{2+} Adsorption Experiments

0.1 g of CMS was added into a conical flask, and then 100 mL of simulated wastewater containing 25, 50, 75, 100, 150, 200, and 250 µg/L of Cd^{2+} were added. The mixtures were stirred under constant temperature of 25 °C at 130 rpm for varying amounts of time. The supernatant was removed and the Cd^{2+} concentration was determined by a flame atomic adsorption spectrometry (FAAS) (WFX-210, Beifen-Ruili Analytical Instrument Co., Beijing, China). The amount of Cd^{2+} removed was calculated by the difference between the initial and equilibrium Cd^{2+} concentrations divided by the solid mass and multiplied by the liquid volume. The Cd^{2+} removal was calculated by the difference between the initial and equilibrium Cd^{2+} concentrations multiplied by 100. All experiments were done in triplicate, and the average and standard deviations are plotted in the figures.

2.4. Material Characterization

The material characterization was conducted using JSM-7500F scanning electron microscope (JEOL, Tokyo, Japan) equipped with energy dispersion spectrum (EDS) system. The BET analysis was performed at 120 °C after 60 min degassing with a N_2 temperature of 77 K. The XRD was performed on a D8 Focus diffractometer from Bruker (Karlsruhe, Germany). Samples were run from 5 to 80° (2θ) with a scanning speed of 2°/min. The thermogravimetric analysis (TGA) was conducted using STA

PT1600 Simultaneous Thermal Analysis manufactured by Linseis (Munich, Germany). The initial mass used was 10 mg, and the heating rate was 10 °C/min. The FTIR spectra were recorded on an IRAffinity FTIR spectrophotometer made by Shimadzu (Tokyo, Japan). Standard KBr pressing method was used for sample preparation and the scan was recorded from 500 to 4000 cm^{-1}.

3. Results and Discussion

3.1. Effects of Different Modification Conditions on Cd^{2+} Removal

The initial chitosan concentrations used for the preparation of CMS was assess first. When the chitosan concentration increased from 0.5% to 2%, the percentage of Cd^{2+} removal from simulated wastewater showed an increasing tendency. Beyond 2%, further increase of chitosan content did not affect Cd^{2+} removal much. At 2% chitosan modification, the Cd^{2+} removal by CMS reached up to 90% (Figure 1).

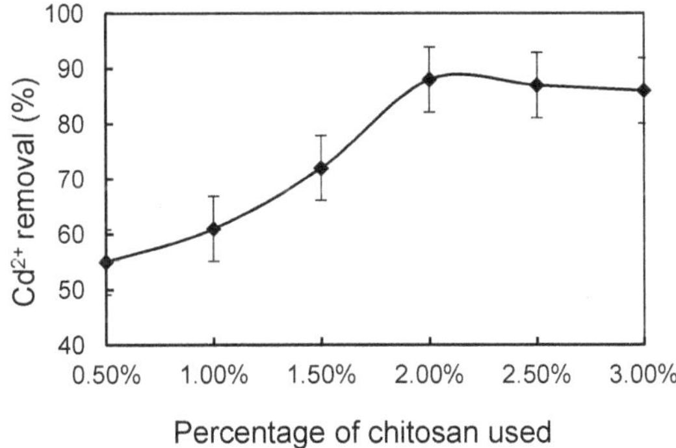

Figure 1. Effect of initial chitosan usage on Cd^{2+} removal.

Chitosan belongs to the natural macromolecule amylose, which has similar properties to chitin and cellulose. When the chitosan concentration was over 2%, a gel-like condition was formed, which limited the contact between chitosan molecules and MS. Therefore, the optimal concentration of chitosan was set at 2% for the fabrication of CMS.

The effect of time of chitosan modification on Cd^{2+} removal was assessed next, with Cd^{2+} removal peaking at 120 min (Figure 2). Thus, for the detailed Cd^{2+} removal experiment, the MS was modified by 2% chitosan for 120 min.

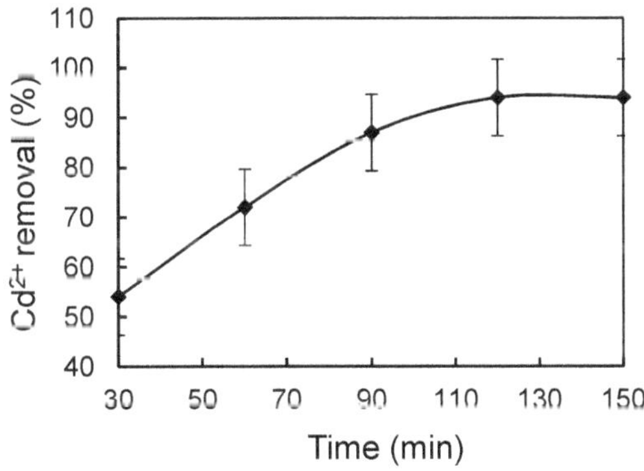

Figure 2. Effect of time of chitosan modification on Cd^{2+} removal.

3.2. Material Characterization of MS and CMS

The MS showed granular particles with smooth surfaces and a particle size of 2–3 µm (Figure 3a). A similar texture was found for the CMS. The SEM images confirmed that the modification did not change the morphology or the internal structure much (Figure 3b). The EDS analyses showed that the MS was mainly composed of O, Na, Al, and Si elements (Figure 3c). The CMS showed an increase in C peak height, as well as the peak at 2.1 keV, attributed to the N element, confirming the uptake of chitosan on the MS (Figure 3d). Although the substitution of Al for Si in tetrahedron resulted in negative charges, the amount of substitution in the MS is relatively small. The BET surface area was 636 and 637 m^2/g, the average pore size was 2.5 and 2.4 nm, and the pore volume was 0.35 and 0.35 cm^3/g for MS and CMS, respectively. The MS results agreed well with a previous study [14]. The uptake of chitosan on MS resulted in a great increase in negative charges. Therefore, its uptake of Cd^{2+} was enhanced.

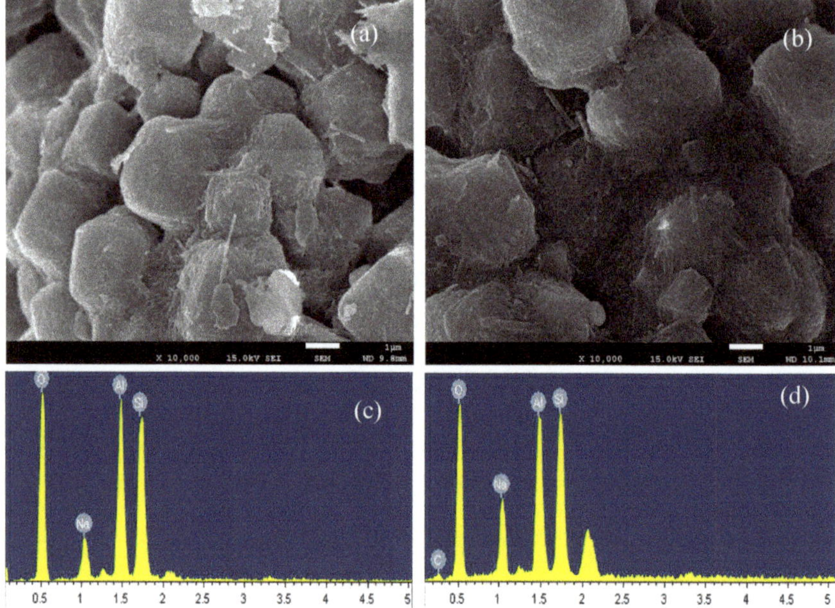

Figure 3. SEM images of MS (**a**) and CMS (**b**) and EDS images of MS (**c**) and CMS (**d**).

The X-ray diffraction (XRD) analyses before and after chitosan modification showed no change in material phases or structure (Figure 4), suggesting that the uptake of chitosan was on the external surfaces. The TGA showed slight different in mass loss (Figure 5). After chitosan modification, there was 2% more mass loss, which could be attributed to mass loss by chitosan. The FTIR analyses showed vibrations at 2850 and 2920 cm^{-1}, attributed to the vibration of chitosan after modification (Figure 6). All these results showed successful modification of MS by chitosan.

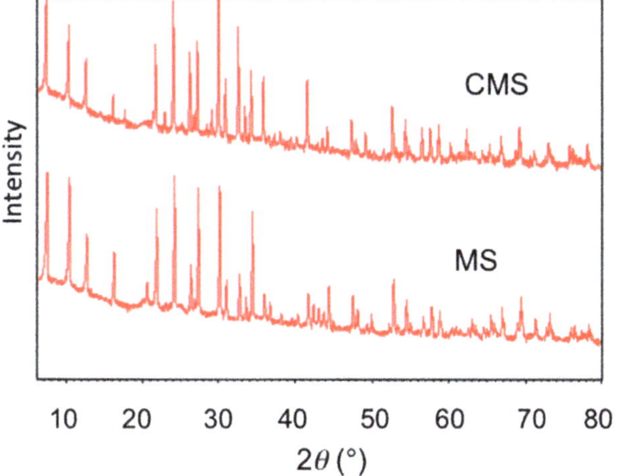

Figure 4. XRD patterns of MS and CMS.

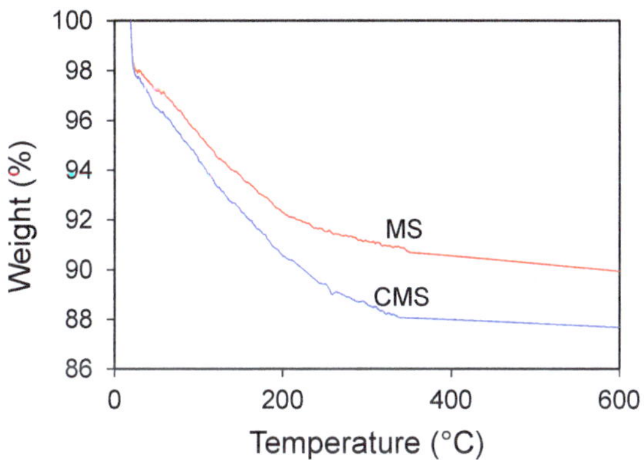

Figure 5. Thermogravimetric analysis of MS and CMS.

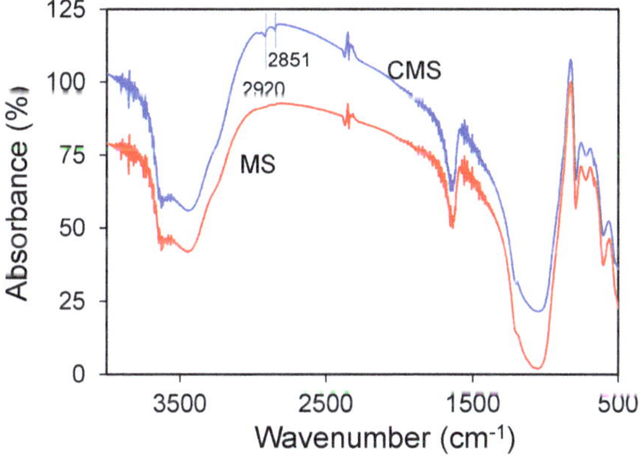

Figure 6. FTIR analysis of MS and CMS.

3.3. Effect of Different Physicochemical Conditions on Cd²⁺ Removal by CMS

With an initial Cd^{2+} concentration of 100 µg/L, the Cd^{2+} uptake and removal varied with equilibrium solution pH (Figure 7). When the solution pH increased from 4 to 7, the Cd^{2+} removal increases quickly from 33.7% to 97.4%. This is mainly because of the fact that, in acid solutions, a high concentration of H ions is in competitive adsorption against Cd^{2+} ions, which impacts the adsorbing effect of the adsorbent. However, with further increase of solution pH beyond 7, the change in Cd^{2+} uptake did not change much (Figure 7). Therefore, the optimal pH value for maximal Cd^{2+} removal was determined to be 7.

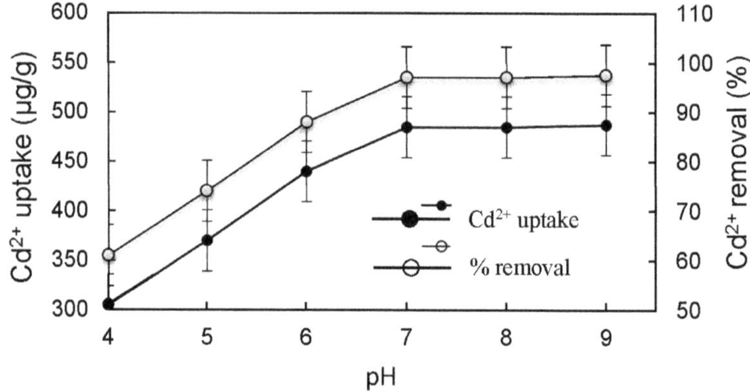

Figure 7. Effect of equilibrium solution pH on Cd^{2+} uptake and removal.

The kinetics of Cd^{2+} removal by CMS of 2% modification is shown in Figure 8. Equilibrium was able to be reached in 90 min. The data were fitted to several kinetic models, and the pseudo second-order model fitted the data best ($r^2 = 0.98$). This has the formula:

$$q_t = \frac{k q_e^2 t}{1 + k q_e t} \tag{1}$$

where k (g/µg-min) is the rate constant of adsorption, q_e (µg/g) the amount of Cd^{2+} adsorbed at equilibrium, and q_t (µg/g) is the amount of Cd^{2+} adsorbed on the surface of the adsorbent at any time, t. The fitted values are $q_e = 625$ µg/g, $k = 4.6 \times 10^{-5}$ g/µg-min, and the rate constant is 17.8 µg/g-min.

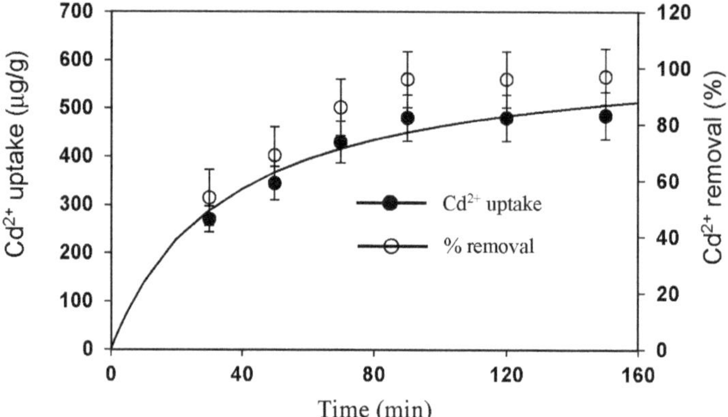

Figure 8. Kinetics of Cd^{2+} uptake and removal.

For the best application, the dosage of CMS was also evaluated. When the dosage of CMS was increased from 0.1 to 0.2 g, the cadmium removal quickly increased from 82.1% to 98.1% (Figure 9). The equilibrium Cd^{2+} was less than 2 µg/L, which meets the standard Cd^{2+} emission of 10 µg/L prescribed in the Standards for Irrigation Water Quality (GB5084-2005) and the MCL and MCLG of 5 µg/L for ground water and drinking water set by USEPA.

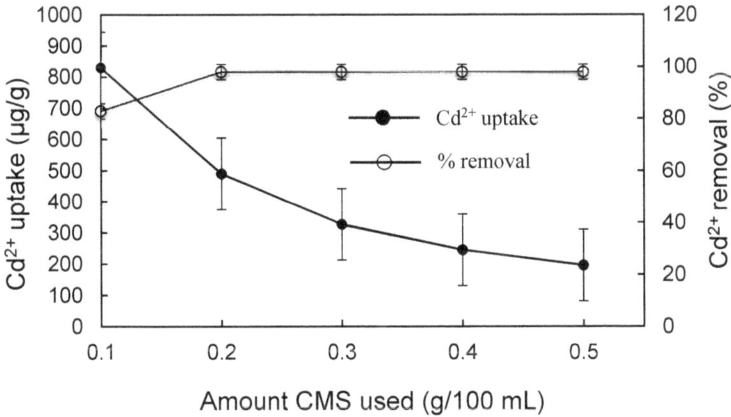

Figure 9. Effect of dosage of CMS on Cd^{2+} uptake and removal.

The isotherm of Cd^{2+} uptake was fitted to both the Langmuir and Freundlich models, and the Langmuir model resulted in a much better fit for the experimental data, with a Cd^{2+} uptake capacity of 1 mg/g (Figure 10 and Table 1). The Langmuir sorption isotherm has the form:

$$C_S = \frac{K_L S_m C_L}{1 + K_L C_L} \tag{2}$$

where C_S is the amount of Cd^{2+} adsorbed at equilibrium (µg/g), S_m the apparent sorption capacity (µg/g), C_L the equilibrium Cd^{2+} concentration (µg/L), and K_L the Langmuir coefficient (L/µg). The Freundlich isotherm has the formula:

$$C_S = K_F C_L^n \tag{3}$$

where K_F, n, and Freundlich constants reflect the adsorption capacity and intensity, respectively. The agreement with the Langmuir model suggested the monolayer adsorption of Cd^{2+} on CMS surfaces, confirming the effectiveness of chitosan modification.

The Cd^{2+} uptake and removal by CMS was compared to that by MS without chitosan modification. A reduction of almost 80 times in equilibrium Cd^{2+} concentration, or a 10-fold increase in Cd^{2+} uptake was achieved by CMS in comparison to MS alone (Table 2).

Preliminary studies showed that the Cd^{2+} uptake and removal was endothermic with the ΔG, ΔH, and ΔS values of 7.4 kJ/mol, 50.6 kJ/mol, and 0.2 kJ/(mol·K), respectively.

The chitosan molecules have large amounts of amino and hydroxyl groups. These functional groups could form complexation with divalent metal cations. Previous results showed that one Cu(II) cation complexed with two amino and groups of chitosan [22]. The agreement of the pseudo-second order kinetic data also pointed to chemisorption [18,19].

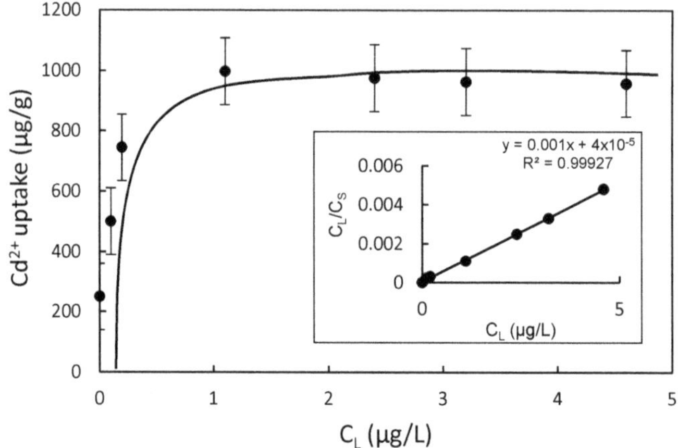

Figure 10. Uptake of Cd^{2+} on CMS. The inset is the linear Langmuir fit to the observed data.

Table 1. Langmuir and Freundlich parameters for Cd^{2+} uptake on CMS.

Langmuir Constant			Freundlich Constant		
S_m (µg/g)	K_L (L/µg)	R^2	logk	$1/n$	R^2
1000	25	0.9993	2.70	0.534	0.47

Table 2. Comparison of Cd^{2+} removal by MS and CMS.

Sample	C_L (µg/L)	Removal (%)	C_S (µg/g)
MS	82.1	17.9	89.5
CMS	1.1	98.7	985

4. Conclusions

1. The optimal condition for the preparation of CMS was mixing 2% chitosan solution with MS for 2 h.
2. SEM and XRD results showed no change in crystal morphology of MS after modification. However, the increase in C and N contents in EDS spectra, the increase in 2% of mass loss in TGA analysis, and the presence of 2850 and 2920 cm^{-1} bands in FTIR analyses confirmed chitosan uptake on MS after modification.
3. The static single factor experiment results showed that under the condition of room temperature, pH of 7, vibration adsorption time of 90 min, and adsorbent dosage of 2 g/L, the Cd^{2+} removal from simulated wastewater with an initial concentration of 100 µg/L was over 96% when using CMS as the adsorbent, which meets the standard Cd^{2+} discharge prescribed in the Standards for Irrigation Water Quality (GB5084-2005) and the MCL and MCLG for groundwater and drinking water standards set by USEPA.
4. The adsorption process of lower concentration of Cd^{2+} in water by CMS fitted with the Langmuir adsorption isotherm model well, with a saturated adsorption capacity of 1 mg/g.

Acknowledgments: This research is supported by the Henan Province Program for Science and Technology Research (172102210055 and 162102210077); the "Twelfth Five Year Plan" 863 project; Key Technology in Remediation of Sewage Irrigation Farmland and Degraded Soil (2012AA101404); Key Scientific Research Project in Henan Province (18B570004 and 15A610008); College Students' Innovative Experimental Program (2016082); and North China University of Water Resources and Electric Power of Engineering Research Fund Project (201610009).

Author Contributions: All authors have contributed in various degrees to the analytical methods used, to the research concept, to the experiment design, to the acquisition of data, or analysis and interpretation of data, to draft the manuscript or to revise it critically for important intellectual content.

Conflicts of Interest: The authors declare no conflict of interest.

References

1. Wang, H.D.; Fang, F.M.; Xie, H.F. Research situation and outlook on heavy metal pollution in water environment of China. *Guangdong Trace Elem. Sci.* **2010**, *17*, 14–18.
2. Zhan, J.; Wei, S.H. Methods of inhibiting cadmium toxicity and its mechanism: A review. *Asian J. Ecotoxicol.* **2012**, *4*, 354–359.
3. Yue, X.Y.; Yi, J. Cadmium poisoning and preventing. *Sichuan Anim. Vet. Sci.* **2000**, *7*, 28–29.
4. Du, L.N.; Yu, R.Z.; Wang, H.Y.; Lu, Y.; Liu, Z.T. Pollution and toxicity of cadmium: A review of recent studies. *J. Environ. Health* **2013**, *2*, 167–174.
5. USEPA. National Primary Drinking Water Regulations. Available online: https://www.epa.gov/ground-water-and-drinking-water/national-primary-drinking-water-regulations (accessed on 15 September 2017).
6. Friberg, L.; Nordberg, G.F.; Vouk, V.B. *Handbook on the Toxicology of Metals*, 3rd ed.; Elsevier: Amsterdam, The Netherlands, 2007; pp. 130–184.
7. World Health Organization; United Nations Environment Programme; Global Environment Monitoring System. *Global Fresh Water Quality*; Published on behalf of the World Health Organization/United Nations Environment Programme; Blackwell Reference: Oxford, UK, 1989.
8. Mustafa, H.T. Cadmium and zinc concentrations in the potable water of the eastern provinces of Saudi Arabia. *Bull. Environ. Contam. Toxicol.* **1988**, *40*, 462–467. [CrossRef] [PubMed]
9. Dai, S.M.; Lv, X.W. Advances on cadmium pollution water treatment technology. *Saf. Environ. Eng.* **2006**, *13*, 63–65.
10. Hu, K.W.; Jia, D.Y.; Zha, C.M. Effect of zeolite on competitive sdsorption of heavy metals ions. *Soil Fertil. Sci. China* **2008**, *3*, 66–69.
11. He, H.P.; Guo, J.G.; Zhu, J.X. An experimental study of adsorption capacity of montmorillonite, kaolinite and illite for heavy metals. *Acta Petrol. Mineral.* **2001**, *20*, 573–578.
12. Zou, Z.H.; He, S.F.; Han, C.Y. Progress of heavy metals liquid waste processing technique. *Technol. Water Treat.* **2010**, *6*, 19–20.
13. Sun, X. *Zeolite NaA and NaX Synthesis from Fly Ash and Their Adsorption of Heavy Metal Ions in Waste Water*; Nan Jing University of Science and Technology: Nanjing, China, 2007.
14. Van Mao, R.L.; Xiao, S.; Ramsaran, A.; Yao, J. Selective removal of silicon from zeolite frameworks using sodium carbonate. *J. Mater. Chem.* **1994**, *4*, 605–610. [CrossRef]
15. Xu, R.R.; Pang, W.Q.; Yu, J.H. *Chemistry-Zeolite and Oorous Materials*; Science Press: Beijing, China, 2004; pp. 356–409.
16. Qin, C.; Li, H.; Xiao, Q.; Liu, Y.; Zhu, J.; Du, Y. Water-solubility of chitosan and its antimicrobial activity. *Carbohydr. Polym.* **2006**, *63*, 367–374. [CrossRef]
17. Hasan, S.; Ghosh, T.; Viswanath, D. Dispersion of chitosan on perlite for enhancement of copper adsorption capacity. *J. Hazard. Mater.* **2008**, *152*, 826–837. [CrossRef] [PubMed]
18. Ngah, W.S.W.; Teong, L.C.; Toh, R.H.; Hanafiah, M.A.K.M. Utilization of chitosan-zeolite composite in the removal of Cu(II) from aqueous solution: Adsorption, desorption and fixed bed column studies. *Chem. Eng. J.* **2012**, *209*, 46–53. [CrossRef]
19. Ngah, W.S.W.; Teong, L.C.; Toh, R.H.; Hanafiah, M.A.K.M. Comparative study on adsorption and desorption of Cu (II) ions by three types of chitosan–zeolite composites. *Chem. Eng. J.* **2013**, *223*, 231–238. [CrossRef]
20. Xie, J.; Li, C.; Chi, L.; Wu, D. Chitosan modified zeolite as a versatile adsorbent for the removal of different pollutants from water. *Fuel* **2013**, *103*, 480–485. [CrossRef]
21. Lu, C.; Yu, S.; Yao, T.; Zeng, C.; Wang, C.; Zhang, L. Zeolite X/chitosan hybrid microspheres and their adsorption properties for Cu (II) ions in aqueous solutions. *J. Porous Mater.* **2015**, *22*, 1255–1263. [CrossRef]
22. Djelad, A.; Morsli, A.; Robitzer, M.; Bengueddach, A.; Di Renzo, F.; Quignard, F. Sorption of Cu (II) ions on chitosan-zeolite X composites: Impact of gelling and drying conditions. *Molecules* **2016**, *21*, 109. [CrossRef] [PubMed]

23. Kołodyńska, D.; Hałas, P.; Franus, M.; Hubicki, Z. Zeolite properties improvement by chitosan modification—Sorption studies. *J. Ind. Eng. Chem.* **2017**, *52*, 187–196. [CrossRef]

24. Hałas, P.; Kołodyńska, D.; Płaza, A.; Gęca, M.; Hubicki, Z. Modified fly ash and zeolites as an effective adsorbent for metal ions from aqueous solution. *Adsorpt. Sci. Technol.* **2017**, *35*, 519–533. [CrossRef]

25. Płaza, A.; Kołodyńska, D.; Hałas, P.; Gęca, M.; Franus, M.; Hubicki, Z. The zeolite modified by chitosan as an adsorbent for environmental applications. *Adsorpt. Sci. Technol.* **2017**. [CrossRef]

26. Hamza, M.F.; Aly, M.M.; Abdel-Rahman, A.A.H.; Ramadan, S.; Raslan, H.; Wang, S.; Vincent, T.; Guibal, E. Functionalization of magnetic chitosan particles for the sorption of U (VI), Cu (II) and Zn (II)—Hydrazide derivative of glycine-grafted chitosan. *Materials* **2017**, *10*, 539. [CrossRef] [PubMed]

Article

Ag-Coated Heterostructures of ZnO-TiO$_2$/Delaminated Montmorillonite as Solar Photocatalysts

Carolina Belver [1],*, Mariana Hinojosa [2], Jorge Bedia [1], Montserrat Tobajas [1], Maria Ariadna Alvarez [1], Vicente Rodríguez-González [2] and Juan Jose Rodriguez [1]

[1] Seccion de Ingenieria Quimica, Facultad de Ciencias, Universidad Autonoma de Madrid, Campus Cantoblanco, E-28049 Madrid, Spain; jorge.bedia@uam.es (J.B.); montserrat.tobajas@uam.es (M.T.); ariadna.alvarez@uam.es (M.A.A.); juanjo.rodriguez@uam.es (J.J.R.)

[2] Division de Materiales Avanzados, IPICYT (Instituto Potosino de Investigación Científica y Tecnológica), Camino a la Presa San José 2055, C.P. 78216 San Luis Potosí, Mexico; mariana.hinojosa.reyes@gmail.com (M.H.); vicente.rdz@ipicyt.edu.mx (V.R.-G.)

* Correspondence: carolina.belver@uam.es; Tel.: +34-91-4978473

Received: 26 July 2017; Accepted: 14 August 2017; Published: 17 August 2017

Abstract: Heterostructures based on ZnO-TiO$_2$/delaminated montmorillonite coated with Ag have been prepared by sol–gel and photoreduction procedures, varying the Ag and ZnO contents. They have been thoroughly characterized by XRD, WDXRF, UV–Vis, and XPS spectroscopies, and N$_2$ adsorption, SEM, and TEM. In all cases, the montmorillonite was effectively delaminated with the formation of TiO$_2$ anatase particles anchored on the clay layer's surface, yielding porous materials with high surface areas. The structural and textural properties of the heterostructures synthesized were unaffected by the ZnO incorporated. The photoreduction led to solids with Ag nanoparticles decorating the surface. These materials were tested as photocatalysts for the degradation of several emerging contaminants with different nitrogen-bearing chemical structures under solar light. The catalysts yielded high rates of disappearance of the starting pollutants and showed quite stable performance upon successive applications.

Keywords: ZnO-TiO$_2$/delaminated montmorillonite; heterostructures; Ag-coating; solar photocatalytic activity; water purification

---

1. Introduction

The development of porous heterostructures based on clay minerals (PCHs) has attracted researchers specially to develop nanoporous materials with predesigned properties for catalytic applications as an alternative to zeolites. In this context, pillared clays (PILCs) were the first studied systems, being prepared by an intercalation of metal-oxopolycations (typically based on aluminum) and subsequently submitted to thermal treatment to consolidate pillars of the metal oxide [1,2]. PILCs have been studied as catalysts because they present high permanent porosity, where the distributed pillars determine, in an ideal perspective, a two-dimensional channel system consisting of micropores comparable to those of zeolites [3]. The possibility of using different types of pillaring agents as well as layered clays of different origins allows for the preparation of PILCs with channels of variable width and pillars of different nature, with potential applications as catalysts in different reactions [4]

Afterwards, new approaches were intended to create other porous clay-derived materials, such as the one reported by Pinnavaia's group called *Porous Clay Heterostructures* (PCHs) [5]. This method is based on a templated synthesis where a surfactant and a cosurfactant are initially intercalated between the clay layers, changing the hydrophobicity of the clay and creating micelles in the interlayer space. Then, a silicon-alkoxide was incorporated and its further hydrolysis and polymerization was controlled to occur around the micelles. Thus, the silica formed was templated by the micelle, generating silica pillars with a very well-ordered pattern. This methodology has been usually employed to prepare mesoporous silica, with modulate and larger porosity than that achievable by common pillaring strategies [6,7]. The interest of this approach has been extended to the preparation of PCHs involving other atoms than Si, such as Al or Ti, which introduce other functional applications related to catalysis and adsorption [8–10].

More recently, a new type of materials related to PILCs and PCHs has been prepared on the basis of oxide nanoparticles (NPs) between delaminated clay sheets, which have been named Delaminated Porous Clay Heterostructures (DPCHs) [11]. This approach uses an organoclay as starting material, which is dispersed in an alcohol, allowing the expansion of the organoclay and the incorporation of alkoxides. Later, the addition of controlled amounts of water provokes the heterocoagulation of the expanded organoclay while the alkoxide is hydrolyzed, giving rise to NPs that remain assembled to the clay network [12,13]. Earlier studies were made on SiO_2- and SiO_2-Al_2O_3 DPCHs [14], but this strategy has been also applied to prepare TiO_2-based DPCHs with photocatalytic applications [15,16].

The investigations on TiO_2-based DPCHs as photocatalysts suggest that the clay can affect the phase of the semiconductor, the size of the NPs formed, and the textural properties of the resulting material. Recent works have demonstrated that the efficiency of these TiO_2-DPCHs photocatalysts can be improved by doping the TiO_2 with transition metals or creating TiO_2-ZnO heterojunctions by a similar synthetic approach that anchors these semiconductors on a delaminated-clay [17–19]. These results confirm that the DPCHs' synthesis can be successfully modulated to obtain photocatalysts active for the degradation of different organic pollutants, with an improved visible absorption capacity and quantum yield. Herein, we report the synthesis of a novel Ag/ZnO-TiO_2/delaminated clay combining the semiconductor properties of the TiO_2-ZnO heterojunctions with the light absorption properties of Ag and the porous texture of the delaminated clay. The main aim is to create photocatalysts with enhanced efficiency towards the degradation of emerging contaminants under solar light.

Based on the literature, several clay-based photocatalysts have been reported using layered montmorillonites and TiO_2 as active phase. Most of them include TiO_2-PILCs with high surface areas, which have demonstrated high decolorization rates for model dyes under UV light. However, these materials have the TiO_2 hindering between the clay layers, being an important drawback for their technological application [20]. In fact, the application of TiO_2-PILCs for the photodegradation of emerging contaminants is not so extended. In this context, different strategies are under study to improve the efficiency of layered clay-based photocatalysts. They are mainly focused on the improvement of the porosity by delaminating the structure, and the anchorage of other semiconductors, such as ZnO, Bi_2O_3, silver halides, and other ternary oxides [20], being necessary to study the relation between the semiconductors and clay materials for the photocatalytic reaction.

In this scenario, the DPCHs appear as a promising way to develop photocatalysts with high efficiency. The identification and removal of these pollutants from water receives nowadays special attention. So far, there is no discharge limitation or regulatory status and their effects on human health and the environment are still under study. Among them, pharmaceuticals, pesticides, or personal care products, which of extended world-wide use are appearing in many aquatic environments as well as in wastewater treatment plants, where they are difficult to remove [21,22]. Different technologies are currently under study for these water pollutants, one of them being photocatalysis. It is based on the ability of a semiconductor material to generate electron–hole pairs induced by the absorption of light with an energy greater than its band gap. These charges can be involved in redox reactions allowing the oxidation of many organic molecules [23,24]. This technology appears as a promising way to remove different pollutants because it opens the chance of using solar light as an energy source. The current work focuses its attention on the degradation of pharmaceuticals (acetaminophen and antipyrine) and pesticide (atrazine) as model emerging contaminants because of their frequent use by the population and their impact on the environment [21].

2. Experimental

2.1. Synthesis of Ag/ZnO-TiO$_2$/Clay Materials

The preparation of Ag/ZnO-TiO$_2$/clay materials follows the heterocoagulation procedure described for TiO$_2$-DPCHs in the literature [16,19]. Summarizing, 1 g of a commercial organo-montmorillonite (Cloisite$^®$ 30B supplied by Southern Clay Products, Gonzales, TX, USA) was dispersed in 10 mL of 2-propanol (Panreac, Castellar del Vallès, Spain) under stirring at 50 °C for 24 h. A solution of titanium (IV) isopropoxide (Sigma Aldrich, St. Louis, MO, USA) in 2-propanol (70% v/v) was slowly added to the clay suspension under stirring, fixing the TiO$_2$/clay weight ratio at 2/1. After 15 min, an aqueous solution of zinc acetate dihydrate (Sigma Aldrich, St. Louis, MO, USA) was added dropwise to the slurry, varying the amount of zinc acetate to obtain ZnO-TiO$_2$ ratios from 0 to 2% (w/w). The mixture was kept at 50 °C under stirring until a gel was formed due to the sol–gel transition of the titanium precursor. The gel was first dried at 60 °C for 24 h and the resulting solid was annealed in air at 500 °C for 4 h (with a 5 °C min^{-1} heating rate), thus removing the organic compounds coming from the organo-montmorillonite and the metal precursors. By this way, several ZnO-TiO$_2$/delaminated clay solids were prepared.

The incorporation of silver particles was carried out by photoreduction [25]. The appropriate amount of the ZnO-TiO$_2$/delaminated clay previously prepared was added to 50 mL of ethanol solution of AgNO$_3$, maintaining magnetic stirring for 30 min to ensure the adsorption of the Ag$^+$ ions on the solid surface. The suspension was further irradiated at 25 °C with a commercial UV lamp (TecnoLite G15T8, 214 nm, 17 W, Jalisco, Mexico) for 1 h under stirring. Afterwards, the solid was separated by filtration and dried for 18 h at 100 °C. The silver amount was adjusted to 1 and 3 wt % Ag. The final delaminated montmorillonite-coated materials were labeled 1C2T-ZnX-AgY, being X the amount of ZnO incorporated (0.5, 1, 2 wt %) and Y the silver amount deposited (1 and 3 wt %). The solid without ZnO (0 wt % used as reference) was named 1C2T-Ag1 following the same label as that in our previous works [16–18].

2.2. Characterization of the Solids

The crystal structure of the samples was analyzed with a Bruker D8 diffractometer (Billerica, MA, USA) equipped with a Sol-X energy dispersive detector to obtain the X-ray diffraction (XRD) patterns. Cu Kα radiation in the 2θ range of 2°–70° with a scanning rate of 1.5° min^{-1} was used. The average crystal size (D) was estimated from the (101) diffraction peak of the anatase phase, the most intense peak, using Scherrer's equation. Wavelength-dispersive X-ray fluorescence spectrometry (WDXRF) was used to determine quantitatively the chemical composition (major and trace elements) of the samples prepared using S8 Tiger Bruker equipment (Billerica, MA, USA). The porous texture of the samples was characterized by N$_2$ adsorption-desorption at −196 °C using a Micromeritics TriStar 123 apparatus (Norcross, GA, USA). The samples were previously outgassed under vacuum at 150 °C for at least 8 h. The total surface area (S$_{BET}$) was quantified by the BET method [26], while the external or nonmicroporous surface area (S$_{EXT}$) and the micropore volume (V$_{MP}$) were estimated using the t-plot method [27]. Finally, the total pore volume (V$_T$) was calculated from the amount of nitrogen (as liquid) adsorbed at a relative pressure of 0.99.

The band gap values were obtained from the UV–vis diffuse reflectance spectra (DRS) carried out on a Shimadzu UV–vis spectrophotometer (model UV-2600, Tokyo, Japan), with an integrating sphere in the 200–900 nm region, using BaSO$_4$ as reference material and the Tauc Plot standard procedure [28]. The representation of $(F(R) \times h\upsilon)^{1/2}$ versus $h\upsilon$ (eV) yields a graph with a linear region whose extrapolation to the x axis gives the band gap value. X-Ray photoelectron spectroscopy (XPS) was used to study the surface composition of the catalysts. The XPS spectra were recorded on a K-Alpha-Thermo Scientific spectrometer (Waltham, MA, USA) using Al Kα X-ray (1486.68 eV) as the excitation source. Binding energies corresponding to Ag 3d, Zn 2p, Ti 2p, and O 1s electrons were obtained using as reference the C 1s line, which was taken as 284.6 eV. The fitting of the XPS signals was made by the least-squares method using peaks with Gaussian–Lorentzian shapes. Scanning electron microscopy (SEM, Hitachi S4800, Tokyo, Japan) using secondary electron (SE) and backscattered electron (BSE) detectors was employed for the analysis of the morphology and particle size of the photocatalysts. The transmission electron microscopy (TEM) images were obtained with a TEM 200 kV, Tecnai G220 from FEI COMPANY (Hillsboro, OR, USA) at an accelerating voltage of 200 kV.

2.3. Photocatalytic Experiments

The photocatalytic degradation tests were performed using antipyrine (phenazone), acetaminophen (paracetamol), or atrazine (Pestanal®) as model compounds. Their respective chemical structures can be seen in Figure S1 of the Electronic Supplementary Information (ESI). The reactions were carried out in 500 mL Pyrex glass reactors (Panreac, Castellar del Vallès, Spain), with two different ports for sampling and flowing air (50 mL min^{-1}), under vigorous magnetic stirring. For the tests, these glass reactors were introduced inside a Suntest solar simulator (Suntest XLS+ photoreactor, ATLAS, Mount Prospect, IL, USA) equipped with a 765–250 W m^{-2} Xe lamp which simulates solar radiation. More details of the photocatalytic reaction system are given elsewhere [16]. In a typical experiment, 250 mg L^{-1} of catalyst (Ag/ZnO-TiO$_2$/clay) were added to 200 mL aqueous solution containing 5 mg L^{-1} of the corresponding target compound. Before the photocatalytic reaction, the solution was stirred in dark overnight to achieve an adsorption equilibrium. After that, the suspension was exposed to solar irradiation for 6 h. The irradiation intensity was fixed at 450 W m^{-2}, and the reaction temperature was monitored to achieve a constant value of 38 ± 1 °C. At given time intervals, 8 mL of the suspension were withdrawn and the photocatalyst was removed by filtration using nylon fiber filters (0.45 μm, Tecknokroma, Sant Cugat del Vallès, Spain). The liquid phase was analyzed by HPLC using a Varian Pro-Start 410 with a UV–vis detector (ProStart 325, Palo Alto, CA, USA) and a reversed phase C18 column (Agilent Technologies, Santa Clara, CA, USA). A mixture of acetonitrile/acetic acid 0.1% *v*/*v* (gradient method: 10/90–40/60% (0–18 min)) was used as the mobile phase, with a constant flow of 0.35 mL min^{-1}. The wavelength used for the detection of each compound was 256, 246, and 270 nm (antipyrine, acetaminophen, and atrazine, respectively).

3. Results and Discussion

3.1. Characterization of ZnO-TiO$_2$/Delaminated Montmorillonite Coated with Ag

When the organo-montmorillonite is dispersed in an alcoholic medium it expands, favoring the intercalation of the titanium precursor. Therefore, the subsequent hydrolysis and condensation upon water addition occur between the clay layers, giving rise to the clay delamination. This procedure has been previously reported for several titania and doped-titania solids [16–18], being a simple and reproducible way to synthesize delaminated porous clay heterostructures. The delamination suffered by the organo-montmorillonite is seen in the X-ray diffractograms depicted in Figure 1. The raw organo-montmorillonite shows a very intense (001) reflection peak at 1.8 nm (characteristic of this organoclay) that disappears after the incorporation of the titania phase (see 1C2T-Ag1 sample). The lack of the (001) reflection comes with the loss of other (001) reflections, indicating that the montmorillonite sheets are disordered in the c-direction, which can be then associated to its delamination. However, at the same time, it maintained the characteristic layered structure, since the other reflections remained unchanged. This trend was maintained when adding ZnO to the synthesis mixture, so that all of the solids prepared can be described as delaminated clay heterostructures.

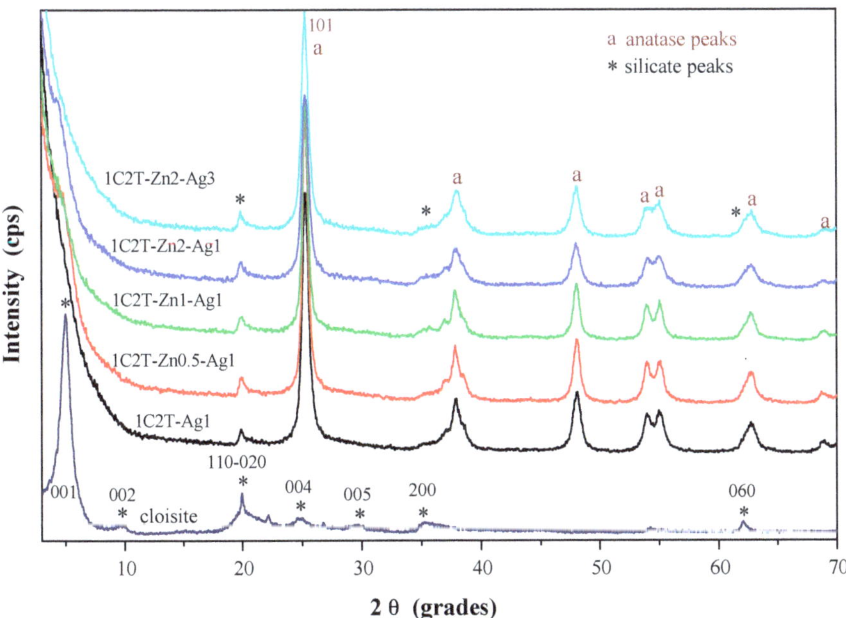

Figure 1. XRD patterns of the starting Cloisite 30B organo-montmorillonite and the synthesized materials.

It is also noteworthy in Figure 1 that all of the samples depict the characteristic reflection peaks of the anatase phase (JCPDS-78-2486), located at 25.5°, 37.8°, 48.1°, 53.9°, 55.2°, 62.9°, and 68.8° 2θ values. There is not any other peak that can be related to the crystallization of other phases, neither from titanium nor from zinc. In addition, the coating with silver particles did not yield any reflection peak typical of Ag^0 crystals (38.1° and 44.2° 2θ values), not even in the solid 1C2T-Zn2-Ag3 with the highest amount of silver. This effect can be associated to the low crystallinity of the silver particles, which is below the detection limits of this technique, and to the high monodispersing of the Ag nanoparticles as will be shown later by TEM characterization. The average crystal size of the anatase phase (D) was estimated by Scherer's equation from the (101) reflection peak. The values are collected in Table 1. With the exception of 1C2T-Zn1-Ag1, the solids yielded similar values (ca. 13 nm), consistent with the anatase size reported previously for the 1C2T heterostructure [16]. This difference can be associated to some small changes in the sol-gel transition caused by the addition of an acid zinc precursor. Usually, the addition of acid during a sol-gel process accelerates the hydrolysis of the alkoxide, changing the subsequent polymerization, which can affect the final crystallization of the titanium oxide [29].

Table 1. Average crystal size of anatase phase (D), chemical composition (wt %), and ZnO/TiO_2 ratio of the solids synthesized.

Sample	D (nm)	TiO_2	ZnO	Ag	ZnO/TiO_2
1C2T-Ag1	12.4	74.30	0.00	1.13	n.d.
1C2T-Zn05-Ag1	13.4	72.70	0.26	0.91	0.36
1C2T-Zn1-Ag1	15.4	71.50	0.49	1.18	0.69
1C2T-Zn2-Ag1	14.0	70.00	1.03	1.17	1.47
1C2T-Zn2-Ag3	13.7	69.80	0.95	2.38	1.36

n.d. not detected.

The delamination of the montmorillonite and anatase crystallization yields to the fixation of the TiO_2 and ZnO, whose percentages are shown in Table 1. The composition of the solids is collected in Table S1 of ESI, where can be seen the SiO_2, Al_2O_3, MgO, and Fe_2O_3 contents characteristic of the original cloisite [14]. As expected, when the amount of ZnO increases, the relative amount of TiO_2 decreases. So, for a better comparison, the ratio between the ZnO and TiO_2 has been calculated and the results are given in Table 1. These values are close to the theoretical contents estimated during the synthesis process, confirming that the methodology used allows the appropriate control of the amount of ZnO and TiO_2 incorporated. In this regard, we should point out that although the presence of ZnO cannot be detected by XRD, because of the low amounts incorporated, the chemical analyses demonstrate that the solids have the desired ZnO contents. All solids have been successfully coated with the expected amount of Ag (1 wt %), except the 1C2T-Zn2-Ag3 that shows 2.4% of Ag instead of the 3% estimated. Since the incorporation of Ag particles first involves a saturation of the solid surface with Ag^+ followed by photoreduction and a final washing step, most probably the surface of the 1C2T-Zn2-Ag3 heterostructure was saturated with the amount fixed and the excess was removed upon the final washing.

Figure 2 represents the N_2 adsorption-desorption isotherms obtained at $-196\ ^\circ C$ of the different samples, including the Cloisite 30B used as starting organoclay. The isotherms have been separated in order to show more clearly the changes of the porous texture upon the different stages of the synthesis. Cloisite 30B shows a type II isotherm of the UIPAC classification, characteristic of nonporous or macroporous solids [30]. It presents an H3 hysteresis loop associated to the nonrigid aggregates of plate-like particles (e.g., clay materials) [29]. The 1C2T heterostructure has been included as a reference. It shows a fairly different porous texture due to the delamination caused by the introduction of TiO_2 between the clay layers. It displays a combination of type I and II isotherms also with an H3 hysteresis loop quite common in layered clay-derived materials such as PILCs and DPCHs [3,31]. This kind of isotherm is typical of a widely distributed porous texture with a contribution of micro-, meso-, and macropores due to the "house-of cards" distribution of the plate-like particles [32,33]. The addition of 1% of Ag (see 1C2T-Ag1 isotherm) results in a slight decrease of the amount of N_2 adsorbed, probably due to a partial pore blockage by Ag particles. Increasing the ZnO content leads to higher amounts of N_2 adsorbed at a low relative pressure, indicative of a higher micropore volume. This trend can be also related to changes on the sol–gel transition because of the acidity of the zinc precursor. Finally, increasing the Ag incorporated from 1 to 3% decreased the amount of N_2 adsorbed. As indicated before, this can be associated to a partial pore blockage by the Ag particles coating on the surface of 1C2T-Zn2.

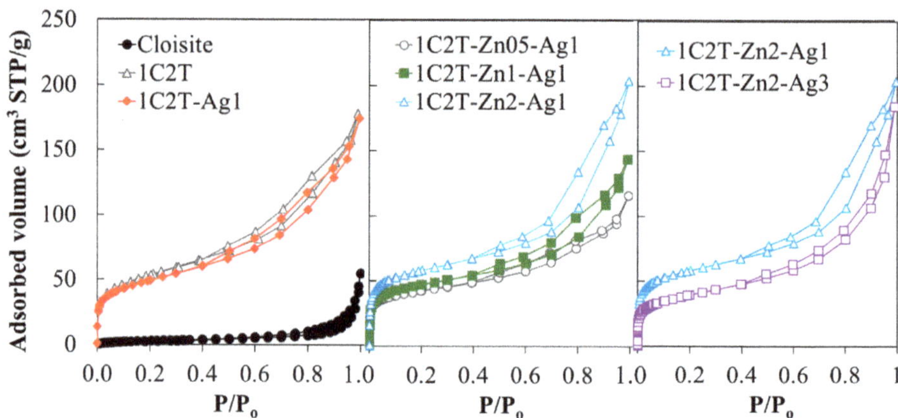

Figure 2. N_2 adsorption-desorption at $-196\ ^\circ C$ of the solids.

Table 2 summarizes the characterization of the porous texture of the solids. It confirms the significant increase of the specific surface area (S_{BET}) of the 1C2T heterostructure with respect to the starting organo-montmorillonite, due to its delamination, which makes accessible both the inner and outer surface of the clay layers [16]. The surface area values are within those previously reported for this kind of DPCH [14,16]. The incorporation of a higher ZnO amount yields a different porous development, as indicated by the increasing values of surface areas and pore volumes (1C2T-Zn0.5-Ag1 < 1C2T-Zn1-Ag1 < 1C2T-Zn2-Ag1). This fact could be associated to the generation of a more disordered porous network, probably related to the different heterocoagulation process occurring upon the sol-gel transition, as indicated before. Regarding Ag incorporation, the values of the surface areas and pore volumes of the corresponding solids confirm the aforementioned partial blockage of porosity by the Ag particles.

Table 2. Characterization of the porous texture and band gap values of the solids.

Sample	S_{BET} (m² g⁻¹)	S_{EXT} (m² g⁻¹)	V_{MP} (cm³ g⁻¹)	V_T (cm³ g⁻¹)	Band Gap (eV)
Cloisite	11	11	n.d.	0.085	n.d.
1C2T	182	152	0.016	0.275	3.21
1C2T-Ag1	173	127	0.021	0.270	3.25
1C2T-Zn0.5-Ag1	144	74	0.033	0.179	3.20
1C2T-Zn1-Ag1	162	83	0.037	0.222	3.20
1C2T-Zn2-Ag1	200	109	0.043	0.316	3.23
1C2T-Zn2-Ag3	138	98	0.018	0.285	3.29

n.d. not detected.

The DRS-UV–visible spectra of the samples (Figure 3A) show that the characteristic band in the UV region is below 360 nm from the charge transference of TiO_2, with the absorption edge around 380 nm [34,35]. These profiles also display an absorption shoulder in the visible region, with a maximum at 480 nm, that could be associated to the surface plasmon absorption characteristic of the Ag^0 particles coating on the solid surface [36]. This absorption shoulder clearly differs according to the ZnO content of the solid, resulting in a sharper absorption in the visible region when the ZnO increases from 0.5 to 2%. This effect suggests a certain type of interaction between the ZnO and the Ag incorporated. Moreover, increasing the silver content from 1 to 3% results in a broader absorption in the visible region, now with the maximum at 510 nm, due to the surface plasmon of the higher silver content. The band gap values (Table 2) were estimated by the Tauc Plot approximation (Figure 3B) considering that these materials are indirect semiconductors as TiO_2 (their main component) [37]. All of the heterostructures yielded band gap values that were very close together, around 3.2–3.3 eV, without important changes due to the incorporation of the ZnO because both semiconductors, TiO_2 and ZnO, are characterized by energy band gap values of 3.2 and 3.2–3.4 eV, respectively. Unlike in some other work [38], here the heterojunction made in these solids between TiO_2 and ZnO did not modify the energy band structure of the final heterostructures.

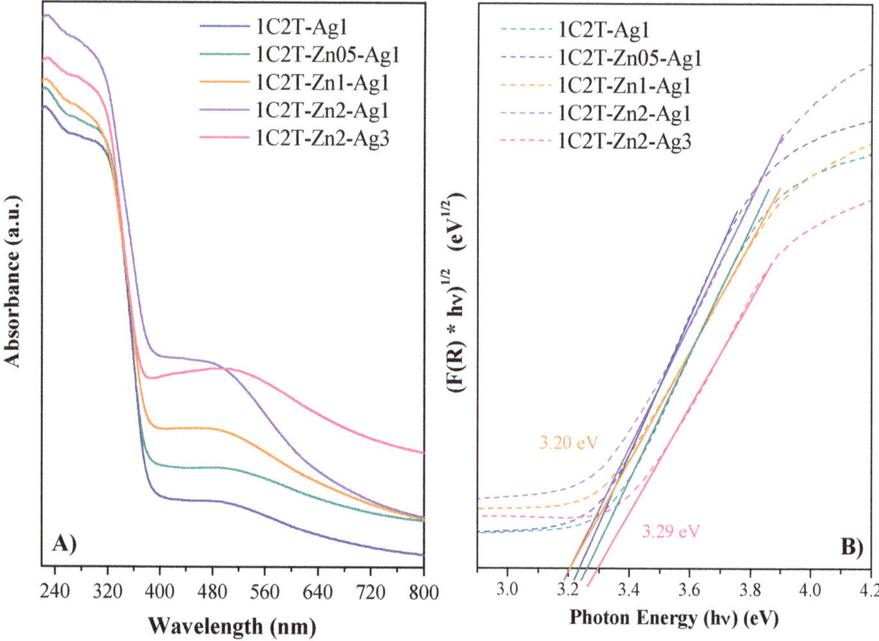

Figure 3. DRS-UV–visible spectra (**A**) and the Tauc Plot (**B**) of the solids.

The surface composition of the heterostructures was studied by XPS. Figure 4 displays the deconvoluted spectra of the Ag 3d, Zn 2p, Ti 2p, and O 1s regions for the 1C2T-Zn0.5-Ag1 and 1C2T-Zn2-Ag3 solids. The Ag 3d region of the XPS spectra of the samples shows a doublet corresponding to Ag $3d_{5/2}$ and Ag $3d_{3/2}$. The Ag $3d_{5/2}$ peaks located at 368.2 and 368.0 eV for 1C2T-Zn0.5-Ag1 and 1C2T-Zn2-Ag3, respectively, are related to the presence of Ag^0, while the Ag $3d_{5/2}$ peaks at 367.4 eV and 367.3 eV for both solids can be attributed to Ag^+ [39–41]. The relative proportions of Ag species of the two solids are given in Table 3, being quite similar for both. The presence of Ag^0 particles corroborates the visible absorption, being more evident in the 1C2T-Zn2-Ag3 solid because of its higher Ag content. The similar Ag^+/Ag^0 ratio is noticeable in spite of the different amounts of Ag (2.4% vs. 1.7% as measured by WDXRF). Although the photoreduction procedure used allows us to coat a solid surface with Ag^0 [24], the photosensitivity of this specie can result in its oxidation to Ag^+, which occurs on the surface in a considerably higher proportion [38]. The deconvolution of the Zn $2p_{3/2}$ profile yielded a peak centered at 1021.4 eV, confirming the presence of Zn^{2+} in both solids [42]. The peak positions of the Ti 2p region, at 464.0 and 458.4 eV, correspond to Ti $2p_{1/2}$ and Ti $2p_{3/2}$, in agreement with the presence of Ti^{4+} [43]. The deconvolution of O 1s confirmed the presence of oxygen in different chemical states [44,45]. Two main bands centered at binding energy values around 429.5 and 431.4 eV were observed in all cases, which can be attributed to Ag_2O and TiO_2 or ZnO, respectively.

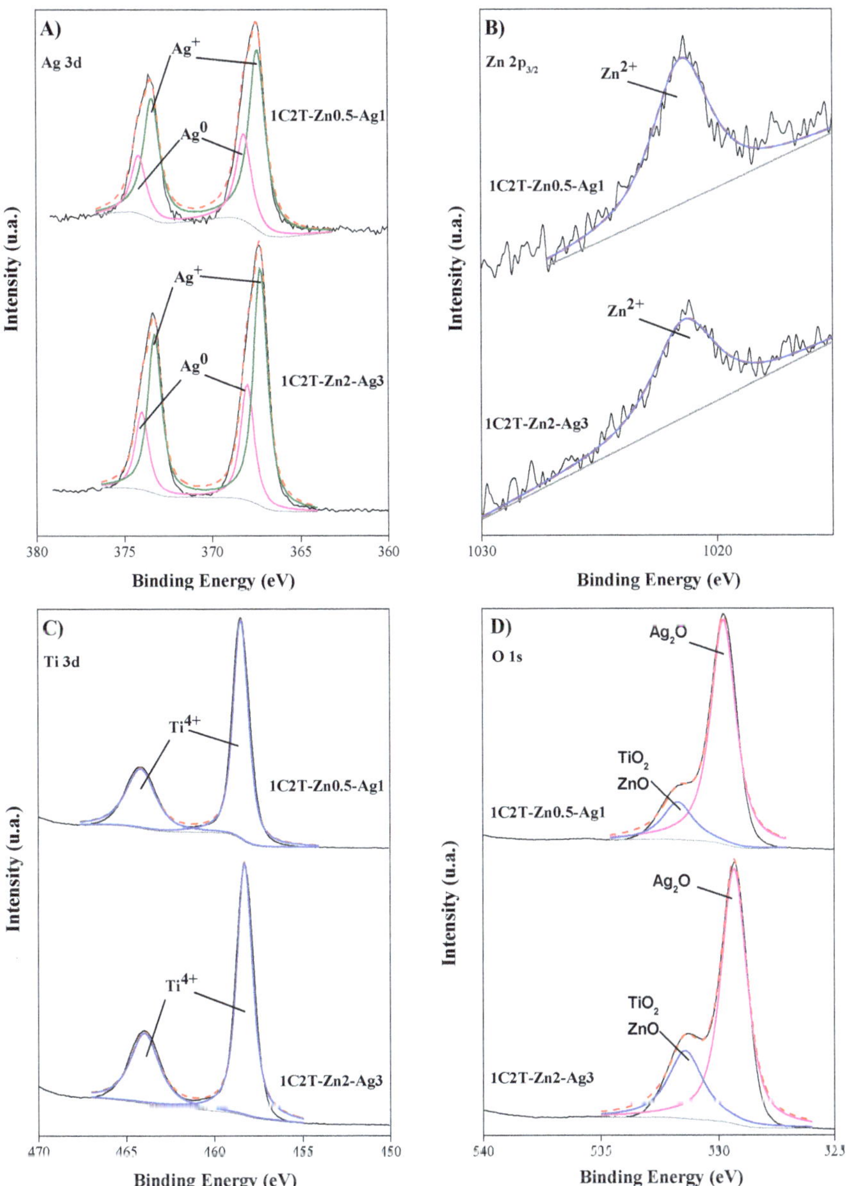

Figure 4. XPS profiles (and their deconvolution) of 1C2T-0.5Zn-1Ag and 1C2T-2Zn-3Ag for: (**A**) Ag; (**B**) Zn; (**C**) Ti; and (**D**) O elements.

Table 3. Surface silver composition (estimated from XPS spectra) of 1C2T-Zn0.5-Ag1 and 1C2T-Zn2-Ag3.

Sample	Ag^+ (%)	Ag^0 (%)	Ag^+/Ag^0
1C2T-Zn0.5-Ag1	69.5	30.5	2.3
1C2T-Zn2-Ag3	68.5	31.5	2.2

Figure 5 shows SEM micrographs of 1C2T-Zn0.5-Ag1 (A and B) and 1C2T-Zn2-Ag3 (C and D) samples observed in secondary (A and C) and back-scattered electrons (B and D). Back-scattered detectors (BSD) provide a much higher mass contrast, therefore the Ag nanoparticles (Ag NPs) appear brighter than the porous support. The images show the presence of the disordered montmorillonite layers, supporting the proper delamination. Furthermore, spongy material can be also observed (Figure 5C) that can be associated to the TiO_2 phase (with its respective lower amount of ZnO) incorporated between the clay layers. Ag nanoparticles can be observed even in the secondary SEM images (Figure 5A), although they are much more clearly observed in the BSD ones.

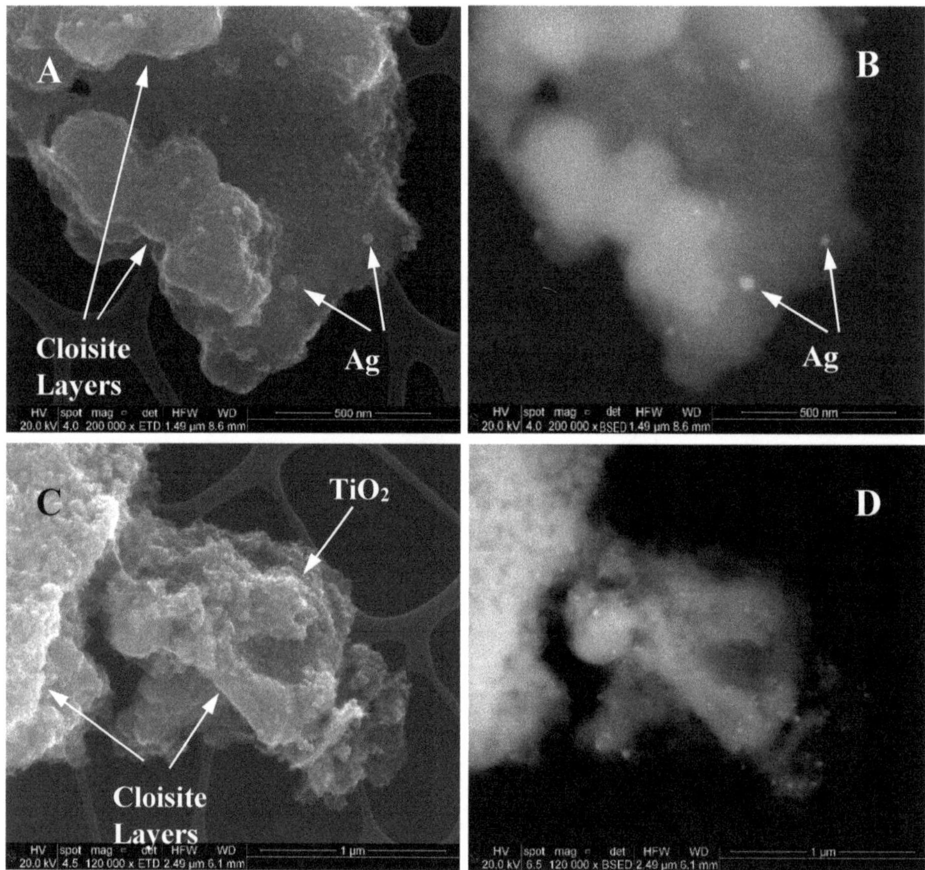

Figure 5. SEM micrographs of 1C2T-Zn0.5-Ag1 (**A,B**) and 1C2T-Zn2-Ag3 (**C,D**), observed in secondary (**A,C**) and back-scattered electrons (**B,D**).

Figure 6 displays some TEM images of 1C2T-Zn0.5-Ag1 (A and B) and 1C2T-Zn2-Ag3 (C and D). Figure 6A shows clearly the presence of different montmorillonite layers surrounded by the TiO_2 phase. These TEM images show also silver nanoparticles (~20 nm) deposited on the TiO_2. Figure 7 depicts an additional TEM image of 1C2T-Zn2-Ag3 and a magnification of a TiO_2 particle (12.6 nm in size), where the lattice space shows a value of approximately 0.35 nm, consistent with that of the anatase phase of TiO_2 [46]. The TiO_2 particles in the samples analyzed show sizes between 10 and 15 nm (Figure 7A). Figure 8 represents the size distribution of the Ag nanoparticles of 1C2T-Zn0.5-Ag1 (A) and 1C2T-Zn2-Ag3 (B). Both distributions are monodisperse, being the larger particles close to

70 nm. The mean Ag particle sizes are 31.3 and 21.8 nm for 1C2T-Zn0.5-Ag1 and of 1C2T-Zn2-Ag3, respectively. Some type of interaction between Ag particles and ZnO could favor the higher dispersion of 1C2T-Zn2-Ag3 interaction, as suggested before, from the radiation absorption in the visible region.

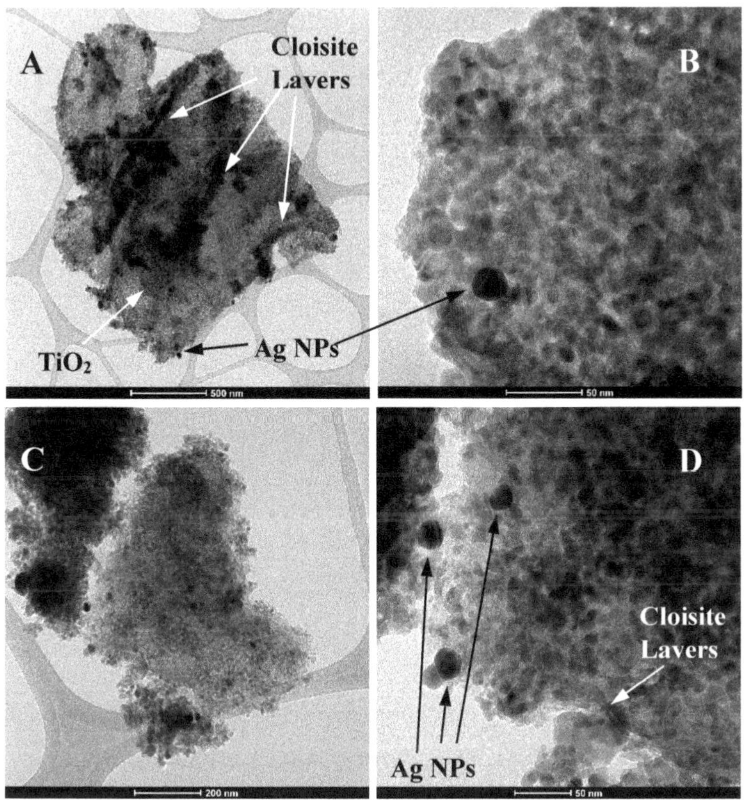

Figure 6. TEM images of 1C2T-Zn0.5-Ag1 (**A,B**) and 1C2T-Zn2-Ag3 (**C,D**). NPs, nanoparticles.

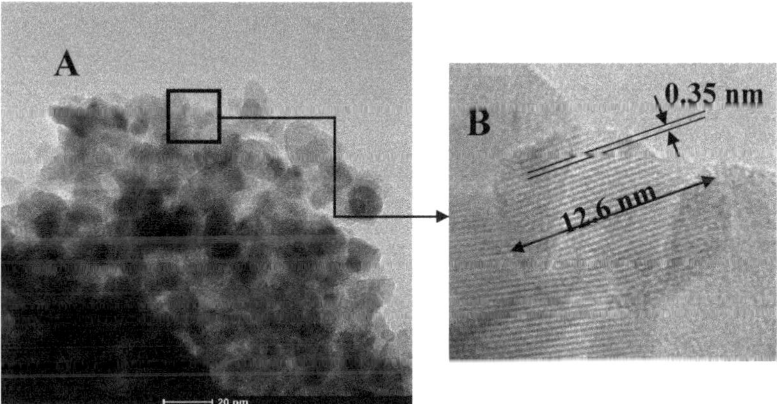

Figure 7. TEM image of 1C2T-Zn2-Ag3 (**A**) and magnification of a TiO_2 particle (**B**).

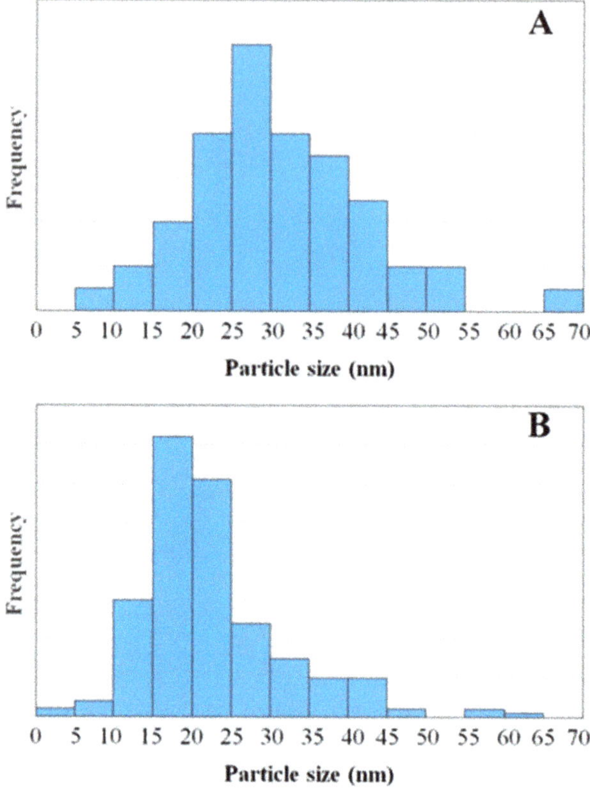

Figure 8. Size distribution of the Ag nanoparticles of 1C2T-Zn0.5-Ag1 (**A**) and 1C2T-Zn2-Ag3 (**B**).

3.2. Photocatalytic Activity

The photocatalytic activity of the delaminated heterostructures was firstly analyzed for the photodegradation of antipyrine (a model emerging pollutant studied in our previous works [18,19]) under solar light. Prior to the photocatalytic experiments, the adsorption capacity of these materials was checked in dark, showing a low antipyrine adsorption. These results were further used to adjust the initial concentration of the target compound in each photocatalytic test. Experiments in the absence of photocatalyst were also carried out, and it was found that noncatalytic photolysis can be neglected, showing the stability of antipyrine (Figure 9) under solar irradiation.

The evolution of the antipyrine concentration upon irradiation time with all of the Ag/ZnO-TiO$_2$ delaminated clay heterostructures is displayed in Figure 9. The photocatalyst 1C2T-Zn0.5-Ag1 degraded nearly 95% of the antipyrine after 360 min, showing higher photocatalytic activity than the 1C2T-Ag1 and the other catalysts with a higher ZnO concentration. The degradation curves were fitted to a pseudo-first-order rate equation, and the resulting values of the kinetic constant (k) are summarized in Table S2 of the ESI. 1C2T-Zn0.5-Ag1 yielded to the highest k value (9.1×10^3 min^{-1}), and higher amounts of ZnO did not improve the photocatalytic efficiency, resulting in lower values from 6.5×10^3 to 7×10^3 min^{-1}. These values were even higher than those described for similar DPCHs without an Ag coating [19]. Regarding the effect of Ag content, 1C2T-Zn2-Ag1 and 1C2T-Zn2-Ag3 showed similar photocatalytic activity, with almost the same rate constant value (6.9×10^3 min^{-1}). Although it was discussed above that the visible absorption of 1C2T-Zn2-Ag3 shifted to higher wavelengths (ca. to 510 nm), that did not cause a positive effect on the photodegradation of antipyrine compared with 1C2T-Zn2-Ag1. Although the characterization of the catalysts indicated that the structural and textural properties of the Ag/ZnO-TiO$_2$ delaminated clay heterostructures were similar, the heterojunction between TiO$_2$-0.5ZnO and the coating with 1 wt % of Ag gave rise to the best photoefficiency, enhancing the separation of the photogenerated charges and avoiding their recombination. Therefore, this photocatalyst and that with the highest ZnO and Ag contents were used for further studies with the other emerging pollutants.

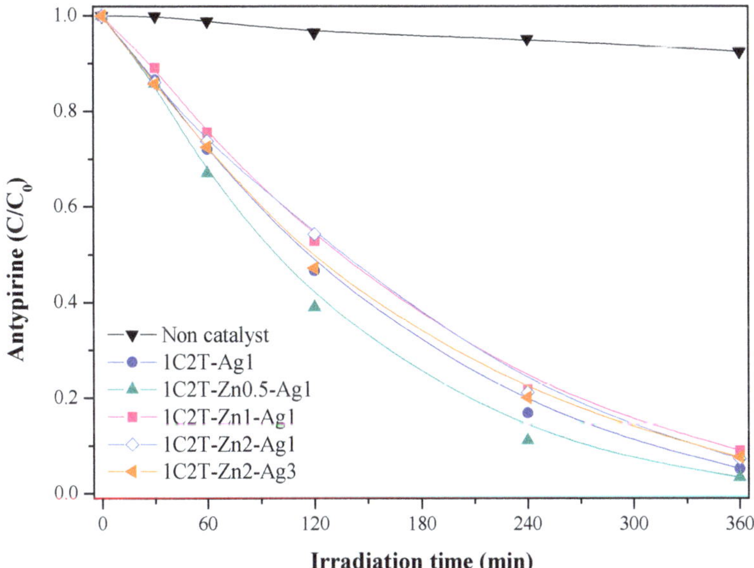

Figure 9. Antipyrine decay upon solar irradiation with the catalysts tested: [Ant]$_0$ = 5 mg L^{-1}; [cat] = 250 mg L^{-1}; W = 450 W m^{-2}; T = 38 $\pm$ 1 °C.

A nitrogenated pharmaceutical compound (acetaminophen) and a nitrochlorinated herbicide (atrazine) were also used to test the photocatalytic activity of the materials tested. The chemical structure of antipyrine is derived from pyrazole, a five-membered aromatic heterocyclic containing two nitrogen atoms in contiguous positions, whereas acetaminophen is a p-aminophenol and atrazine contains a chlorinated s-triazine ring. The stability under solar irradiation was firstly checked without photocatalyst (Figure 10), confirming the absence of noncatalytic photolysis. Furthermore, the adsorption capacity of the catalysts was checked for 18 h in dark, showing a low adsorption towards the two contaminants (<5% for acetaminophen and 14% for atrazine). Figure 10 displays the evolution of acetaminophen and atrazine concentrations upon reaction time with the catalysts tested. It can be seen that the highest photocatalytic activity belongs to 1C2T-Zn0.5-Ag1, reaching 92 and 89% degradation of atrazine and acetaminophen after 240 min, and around 96% and complete conversion after 360 min, respectively. Similar photocatalytic activity was found for antipyrine degradation with this catalyst, resulting in similar rate constants for the three contaminants (Table S2 of the ESI). Therefore, the nitrogenated ring structure of the compounds does not cause any noticeable effect on the 1C2T-Zn0.5-Ag1's photoefficiency under solar light. Nevertheless, the lower activity of the 1C2T-Zn2-Ag3 photocatalyst made clearer the effect of the chemical structure of the contaminants, resulting in the following sequence of degradation rates: antipyrine > acetaminophen > atrazine. This suggests that the pyrazole structure of antipyrine found it easier to interact with the radicals generated during the photocatalytic reaction than did the aminophenol structure of acetaminophen, and much more than the s-triazine ring structure of atrazine. Thus, the influence of the nitrogenated ring structure on the efficiency of the photocatalytic process is noteworthy, but only when a less active photocatalyst is used, since the 1C2T-Zn0.5-Ag1 heterostructure performed efficiently as photocatalyst with the three target compounds tested.

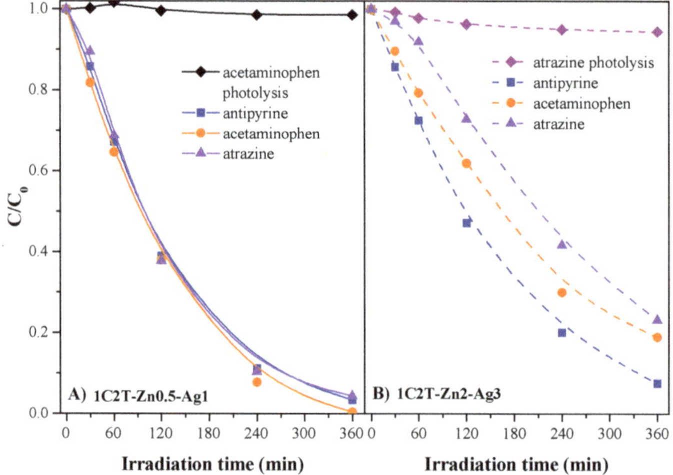

Figure 10. Evolution of the concentration of antipyrine, acetaminophen, and atrazine upon solar irradiation time with 1C2T-Zn0.5-Ag1 and 1C2T-Zn2-Ag3.

A complementary study was performed to learn about the reusability of the Ag/ZnO-TiO$_2$ delaminated clay heterostructures that were synthetized. With this aim, a sequence of three cycles of photocatalytic degradation under solar irradiation during 6 h were carried out using acetaminophen as the target compound and the most active photocatalyst (1C2T-Zn0.5-Ag1). The operating conditions were maintained as in the previously described experiments, and before each run the catalyst was thoroughly washed with water and dried for 1 h at 60 °C. Figure 11 shows the photodegradation percentages reached in the three successive cycles. The photocatalytic activity remained almost unchanged, reaching about 90% degradation. These materials have also the advantage that they can be easily removed from the aqueous solution, showing a fast decantation rate compared with the conventional photocatalysts used as powders.

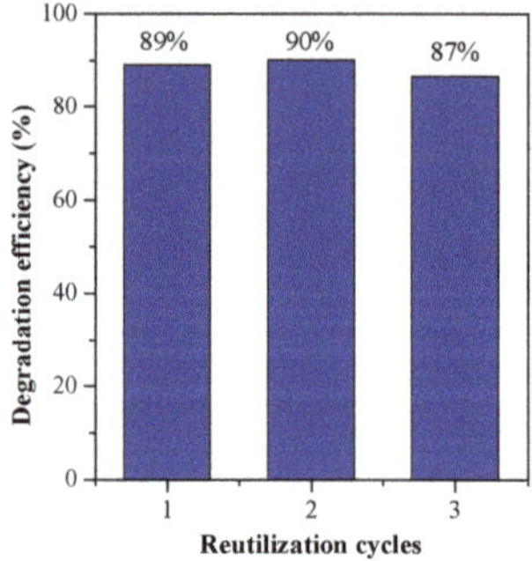

Figure 11. Reutilization test of 1C2T-Zn0.5-Ag1 for the photocatalytic degradation of acetaminophen during three successive runs.

4. Conclusions

Novel heterostructures based on a ZnO-TiO$_2$/delaminated montmorillonite coated with Ag nanoparticles, with different ZnO and Ag loads, were successfully synthesized by sol-gel and further photoreduction. The structural characterization revealed, in all cases, the delamination of the montmorillonite and the crystallization of TiO$_2$ anatase anchored on the surface of the clay layers. The fixation of ZnO and Ag was confirmed by chemical analyses and XPS, although without the formation of any crystallized phase. The resulting heterostructures exhibited high surface areas with a contribution from both micro- and mesopores. The surface characterization revealed the presence of silver as both Ag$^+$ and Ag^0 at surface concentrations, and the relative amounts were unchanged regardless of the total amount coated. The presence of silver nanoparticles induced the light absorption of the solids in the visible region. These Ag particles displayed an average size of 25–30 nm, showing monodisperse size distributions. All of these heterostructures were effective for the photodegradation of antipyrine under solar light, being 1C2T-Zn0.5-Ag1 the most active. Its efficiency was similar for the three different chemical structures tested as model emerging contaminants (antipyrine, acetaminophen, and atrazine, this last containing also chlorine). The most active catalyst (1C2T-Zn0.5-Ag1) was tested in three successive runs, showing almost unaltered performance.

Supplementary Materials: The following are available online at www.mdpi.com/1996-1944/10/8/960/s1. The supplementary material includes: Table S1: Chemical analyses (wt %) of heterostructures, referred to ignited solids (0% water); Table S2: Values of the rate constant for antipyrine, acetaminophen, and atrazine degradation for the Ag/ZnO-TiO$_2$ delaminated clay heterostructures; Figure S1: Chemical structure of the pharmaceuticals and herbicide used as model of emerging contaminants.

Acknowledgments: The authors acknowledge the financial support from Spanish MINECO (project CTQ2016-78576-R). Carolina Belver is indebted to the MINECO for a Ramon y Cajal postdoctoral contract. Thanks to Gamarra and the SAIUEx service for the WDXRF, XPS, and TEM characterization.

Author Contributions: Carolina Belver conceived and designed the experiments; Carolina Belver, Mariana Hinojosa, Jorge Bedia, Montserrat Tobajas, and Maria Ariadna Alvarez performed the experiments; Carolina Belver, Jorge Bedia, Montserrat Tobajas, and Maria Ariadna Alvarez analyzed the data and wrote the manuscript; Mariana Hinojosa and Vicente Rodríguez-González contributed reagents/materials/synthesis tools; and Juan Jose Rodriguez and Vicente Rodríguez-González revised/discussed the paper.

References

1. Brindley, G.W.; Sempels, R.E. Preparation and properties of some hydroxy-aluminum beidellites. *Clay Miner.* **1977**, *12*, 229–237. [CrossRef]
2. Lahav, N.; Shani, U.; Shabtai, J. Cross-linked smectites. I. Synthesis and properties of hydroxy-aluminum montmorillonite. *Clays Clay Miner.* **1978**, *26*, 107–115. [CrossRef]
3. Gil, A.; Korili, S.A.; Vicente, M.A. Recent advances in the control and characterization of the porous structure of pillared clay catalysts. *Catal. Rev.* **2008**, *50*, 153–221. [CrossRef]
4. Gil, A.; Korili, S.A.; Trujillano, R.; Vicente, M.A. *Pillared Clays and Related Catalysts*, 1st ed.; Springer: New York, NY, USA, 2010; ISBN 978-1-44196670-4.
5. Galarneau, A.; Barodawalla, A.; Pinnavaia, T.J. Porous clay heterostructures formed by gallery-templated synthesis. *Nature* **1995**, *374*, 529–531. [CrossRef]
6. Galarneau, A.; Barodawalla, A.; Pinnavaia, T.J. Porous clay heterostructures (PCH) as acid catalysts. *Chem. Commun.* **1997**, 1661–1662. [CrossRef]
7. Polverejan, M.; Pauly, T.R.; Pinnavaia, T.J. Acidic porous clay heterostructures (PCH). Intragallery assembly of mesoporous silica in synthetic saponite clays. *Chem. Mater.* **2000**, *12*, 2698–2704. [CrossRef]
8. Chmielarz, L.; Kustrowski, P.; Dziembaj, R.; Cool, P.; Vansant, E.F. Selective catalytic reduction of NO with ammonia over porous clay heterostructures modified with copper and iron species. *Catal. Today* **2007**, *119*, 181–186. [CrossRef]

9. Tchinda, A.J.; Ngameni, E.; Kenfack, I.T.; Walkarius, A. One-Step preparation of thiol-functionalized porous clay heterostructures: Application to Hg(II) binding and characterization of mass transport issues. *Chem. Mater.* **2009**, *21*, 4111–4121. [CrossRef]

10. Qu, F.; Zhu, L.; Yang, K. Adsorption behaviors of volatile organic compounds (VOCs) on porous clay heterostructures (PCH). *J. Hazard. Mater.* **2009**, *170*, 7–12. [CrossRef] [PubMed]

11. Aranda, P.; Belver, C.; Ruiz-Hitzky, E. Inorganic heterostructured materials based on clay mineral. In *Materials and Clay Minerals*, 1st ed.; Drummy, L.F., Ogawa, M., Aranda, P., Eds.; CMS Workshop Lectures; The Clay Minerals Society: Chantilly, VA, USA, 2013; Volume 18, pp. 21–40. [CrossRef]

12. Letaïef, S.; Ruiz-Hitzky, E. Silica-clay nanocomposites. *Chem. Commun.* **2003**, 2996–2997. [CrossRef]

13. Letaïef, S.; Martín-Luengo, M.A.; Aranda, P.; Ruiz-Hitky, E. A colloidal route for delamination of layered solids: Novel porous-clay nanocomposites. *Adv. Funct. Mater.* **2006**, *16*, 401–409. [CrossRef]

14. Belver, C.; Aranda, P.; Martín-Luengo, M.A.; Ruiz-Hitzky, E. New silica/alumina-clay heterostructures: Properties as acid catalysts. *Microporous Mesoporous Mater.* **2012**, *147*, 157–166. [CrossRef]

15. Manova, E.; Aranda, P.; Martín-Luengo, M.A.; Letaief, S.; Ruiz-Hitzky, E. New titania-clay nanostructured porous materials. *Microporous Mesoporous Mater.* **2010**, *131*, 252–260. [CrossRef]

16. Belver, C.; Bedia, J.; Rodriguez, J.J. Titania-clay heterostructures with solar photocatalytic applications. *Appl. Catal. B Environ.* **2015**, *176–177*, 278–287. [CrossRef]

17. Belver, C.; Bedia, J.; Álvarez-Montero, M.A.; Rodriguez, J.J. Solar photocatalytic purification of water with Ce-doped TiO_2/clay heterostructures. *Catal. Today* **2016**, *266*, 36–45. [CrossRef]

18. Belver, C.; Bedia, J.; Rodriguez, J.J. Zr-doped TiO_2 supported on delaminated clay materials for solar photocatalytic treatment of emerging pollutants. *J. Hazard. Mater.* **2017**, *322*, 233–242. [CrossRef] [PubMed]

19. Tobajas, M.; Belver, C.; Rodriguez, J.J. Degradation of emerging pollutants in water under solar irradiation using novel TiO_2-ZnO/clay nanoarchitectures. *Chem. Eng. J.* **2017**, *309*, 596–606. [CrossRef]

20. Liu, J.; Zhang, G. Recent advances in synthesis and applications of clay-based photocatalysts: A review. *Phys. Chem. Chem. Phys.* **2014**, *16*, 8178–8192. [CrossRef] [PubMed]

21. Richardson, S.D.; Ternes, T.A. Water analysis: Emerging contaminants and current issues. *Anal. Chem.* **2011**, *83*, 4614–4648. [CrossRef] [PubMed]

22. Pal, A.; Gin, K.Y.-H.; Lin, A.Y.-C.; Reinhard, M. Impacts of emerging contaminants on freshwater resources: Review of recent occurrences, sources, fate and effects. *Sci. Total Environ.* **2010**, *408*, 6062–6069. [CrossRef] [PubMed]

23. Hoffmann, M.R.; Martin, S.T.; Choi, W.; Bahnemann, D.W. Environmental applications of semiconductor photocatalysis. *Chem. Rev.* **1995**, *95*, 69–96. [CrossRef]

24. Mills, A.; Le Hunte, S. An overview of semiconductor photocatalysis. *J. Photochem. Photobiol. A* **1997**, *108*, 1–35. [CrossRef]

25. Hernández-Gordillo, A.; Rodríguez-González, V. Silver nanoparticles loaded on Cu-doped TiO_2 for the effective reduction of nitro-aromatic contaminants. *Chem. Eng. J.* **2015**, *261*, 53–59. [CrossRef]

26. Brunauer, S.; Emmett, P.H.; Teller, E. Adsorption of Gases in Multimolecular Layers. *J. Am. Chem. Soc.* **1938**, *60*, 309–319. [CrossRef]

27. Lippens, B.C.; de Boer, J.H. Studies on pore systems in catalysts: V. The *t* method. *J. Catal.* **1965**, *4*, 319–323. [CrossRef]

28. Tauc, J. Absorption edge and internal electric fields in amorphous semiconductors. *Mater. Res. Bull.* **1970**, *5*, 721–726. [CrossRef]

29. Turova, N.Y.; Turevskaya, E.P.; Kessler, V.G.; Yanovskaya, M.I. *The Chemistry of Metal Alkoxides*; Kluwer Academic Publishers: New York, NY, USA, 2002; ISBN 0-7923-7521-1.

30. Thommes, M.; Kaneko, K.; Neimark, A.V.; Olivier, J.P.; Rodriguez-Reinoso, F.; Rouquerol, J.; Sing, K.S.W. Physisorption of gases, with special reference to the evaluation of surface area and pore size distribution (IUPAC Technical Report). *Pure Appl. Chem.* **2015**, *87*, 1051–1069. [CrossRef]

31. Ruiz-Hitzky, E.; Aranda, P.; Belver, C. Nanoarchitectures based on clay materials. In *Manipulation on Nanoscale Materials: An Introduction to Nanoarchitectonics*; Ariga, K., Ed.; The Royal Society of Chemistry: Cambridge, UK, 2012; pp. 87–111, ISBN 978-1-84973-415-8.

32. Li, K.; Lei, J.; Yuan, G.; Weerachanchai, P.; Wang, J.-Y.; Zhao, J.; Yang, Y. Fe-, Ti-, Zr- and Al-pillared clays for efficient catalytic pyrolysis of mixed plastics. *Chem. Eng. J.* **2017**, *317*, 800–809. [CrossRef]

33. Belessi, V.; Lambropoulou, D.; Konstantinou, I.; Katsoulidis, A.; Ponomis, P.; Petridis, D.; Albanis, T. Structure and photocatalytic performance of TiO_2/clay nanocomposites for the degradation of dimethachlor. *Appl. Catal. B Environ.* **2007**, *73*, 292–299. [CrossRef]

34. Pelaez, M.; Nolan, N.T.; Pillai, S.C.; Seery, M.K.; Falaras, P.; Kontos, A.G.; Dunlop, P.S.M.; Hamilton, J.W.J.; Byrne, J.A.; O'shea, K.; et al. A review on the visible light active titanium dioxide photocatalysts for environmental applications. *Appl. Catal. B* **2012**, *125*, 331–349. [CrossRef]

35. Chen, J.; Qiu, F.; Xu, W.; Cao, S.; Zhu, H. Recent progress in enhancing photocatalytic efficiency of TiO_2-based materials. *Appl. Catal. A* **2015**, *495*, 131–140. [CrossRef]

36. Holtz, R.D.; Souza-Filho, A.G.; Brocchi, M.; Martins, D.; Durán, N.; Alves, O.L. Development of nanostructured silver vanadates decorated with silver nanoparticles as a novel antibacterial agent. *Nanotechnology* **2010**, *21*, 185102–185110. [CrossRef] [PubMed]

37. Kwiatkowski, M.; Bezberkhyy, I.; Skompska, M. ZnO nanorods covered with a TiO_2 layer: Simple sol-gel preparation, and optical, photocatalytic and photoelectrochemical properties. *J. Mater. Chem. A* **2015**, *2*, 12748–12760. [CrossRef]

38. Li, D.; Jiang, X.; Zhang, Y.; Zhang, B. A novel route to ZnO/TiO_2 heterojunction composite fibers. *J. Mater. Res.* **2013**, *28*, 507–512. [CrossRef]

39. Belver, C.; Adan, C.; García-Rodríguez, S.; Fernández-García, M. Photocatalytic behavior of silver vanadates: Microemulsion synthesis and post-reaction characterization. *Chem. Eng. J.* **2013**, *224*, 24–31. [CrossRef]

40. Mangayayam, M.; Kiwi, J.; Giannakis, S.; Pulgarin, C.; Zivkovic, I.; Magrez, A.; Rtimi, S. FeOx magnetization enhancing *E. coli* inactivation by orders of magnitude on Ag-TiO_2 nanotubes under sunlight. *Appl. Catal. B* **2017**, *202*, 438–445. [CrossRef]

41. Deng, X.; Li, M.; Zhang, J.; Hu, X.; Zheng, J.; Zhang, N.; Chen, B.H. Constructing nano-structure on silver/ceria-zirconia towards highly active and stable catalyst for soot oxidation. *Chem. Eng. J.* **2017**, *313*, 544–555. [CrossRef]

42. Pirhashemi, M.; Habibi-Yangjeh, A. Ultrasonic-assisted preparation of plasmonic ZnO/Ag/Ag_2WO_4 nanocomposites with high visible-light photocatalytic performance for degradation of organic pollutants. *J. Colloid Interface Sci.* **2017**, *491*, 216–229. [CrossRef] [PubMed]

43. Xu, H.; Liao, J.; Yuan, S.; Zhao, Y.; Zhang, M.; Wang, Z.; Shi, L. Tuning the morphology, stability and photocatalytic activity of TiO_2 nanocrystal colloids by tungsten doping. *Mater. Res. Bull.* **2014**, *51*, 326–331. [CrossRef]

44. Hoflund, G.B.; Hazos, Z.F.; Salaita, G.N. Surface characterization study of Ag, AgO, and Ag_2O using X-ray photoelectron spectroscopy and electron energy-loss spectroscopy. *Phys. Rev. B* **2000**, *62*, 11126–11133. [CrossRef]

45. Lee, A.Y.; Blakeslee, D.M.; Powell, C.J.; Rumble, J.R. Development of the web-based NIST X-ray Photoelectron Spectroscopy (XPS) Database. *Data Sci. J.* **2002**, *1*, 1–12. [CrossRef]

46. Tahir, M.; Amin, N.S. Photocatalytic reduction of carbon dioxide with water vapors over montmorillonite modified TiO_2 nanocomposites. *Appl. Catal. B* **2013**, *142–143*, 512–522. [CrossRef]

 *materials*

Article

Study of CeO$_2$ Modified AlNi Mixed Pillared Clays Supported Palladium Catalysts for Benzene Adsorption/Desorption-Catalytic Combustion

Jingrong Li, Shufeng Zuo * , Peng Yang and Chenze Qi

Zhejiang Key Laboratory of Alternative Technologies for Fine Chemicals Process, Shaoxing University, Shaoxing 312000, China; sfzuo@126.com (J.L.); yangpeng1898@163.com (P.Y.); qichenze@usx.edu.cn (C.Q.)
* Correspondence: sfzuo@usx.edu.cn; Tel.: +86-571-8834-5683

Received: 5 July 2017; Accepted: 11 August 2017; Published: 15 August 2017

Abstract: A new functional AlNi-pillared clays (AlNi-PILC) with a large surface area and pore volume was synthesized. The performance of adsorption/desorption-catalytic combustion over CeO$_2$-modified Pd/AlNi-PILC catalysts was also studied. The results showed that the d_{001}-value and specific surface area (S_{BET}) of AlNi-PILC reached 2.11 nm and 374.8 m^2/g, respectively. The large S_{BET} and the d_{001}-value improved the high capacity for benzene adsorption. Also, the strong interaction between PdCe mixed oxides and AlNi-PILC led to the high dispersion of PdO and CeO$_2$ on the support, which was responsible for the high catalytic performance. Especially, 0.2% Pd/12.5% Ce/AlNi-PILC presented high performance for benzene combustion at 240 °C and high CO$_2$ selectivity. Also, the combustion temperatures were lower compared to the desorption temperatures, which demonstrated that it could accomplish benzene combustion during the desorption process. Furthermore, its activity did not decrease after continuous reaction for 1000 h in dry air, and it also displayed good resistance to water and the chlorinated compound, making it a promising catalytic material for the elimination of volatile organic compounds.

Keywords: AlNi-PILC; Pd-Ce; catalytic combustion; benzene; TPD/TPSR

1. Introduction

Volatile organic compounds (VOCs) have high vapor pressure and low water solubility at room temperature, and already have been recognized as major contributors to air pollution. They mainly come from industrial processes, fossil fuel combustion, cement concrete, and furniture coatings [1]. Among various VOCs, the carcinogenic benzene is one of the most abundant found in either industrial operations or at home [2]. It can bring photo-chemical smog, ozone generation, and offensive odors. The catalytic combustion method has been proved to be highly-efficient for VOCs degradation, providing carbon dioxide and water as final products (because of its higher efficiency, lower operating temperature, and less harmful by-products than thermal oxidation [3–7]).

The studies of catalysts for VOCs catalytic combustion have been reported, focusing on three types of catalysts based on noble metals [8,9], transition metal oxides [10,11] and rare earth metal oxides [10]. Generally, noble metal catalysts (Pt, Pd, and Au) [12–14] are commonly used for VOCs oxidation, and they usually represent higher activity than transition metal oxides. Particularly, supported Pd catalyst is one of the most used materials, due to its high activity for deep oxidation of VOCs at relatively low temperatures [15–22]. Moreover, as important promoters, rare earth elements (REE) with a special electronic structure have drawn much attention in recent years. REE can decrease the amount of noble metals, stabilize supports against thermal sintering, improve the performance of catalysts in storing/releasing oxygen, and reduce the reaction activation energy [23–27].

As is well known, the support is also an important factor for the performance of supported noble metal catalysts, and the choice of catalyst support usually depends on its specific surface area (S_{BET}), pore size, and the capacity for interaction with metals. Generally, higher S_{BET} can provide more active sites and larger pore size more easily and allows the reactants to approach those catalytic active sites. Montmorillonite KSF (MMT) has been applied in initial clay due to its stable structure, low cost, and environmental compatibility [28–31]. Notably, as the modification of MMT, pillared clays (PILC) have a large S_{BET} and pore volume (V_P), and the porous structure and physicochemical properties of MMT are improved significantly. In order to further improve the PILC performance, more attention has been directed toward PILC with mixed oxide pillars including Al-Zr, Al-Fe, and Al-Cr-PILC [6,9,32]. However, their applications are limited due to their poor thermal stability and durability. Some reports illustrate that various Ni-containing porous materials have good thermal and hydrothermal stability, such as Ni-Al-MCM-41 [33–35], Ni-zeolite [36], and Ni-Al-SBA-51 [37–39]. However, disadvantages still remain, including the complicated preparation process for supports. Therefore, there is an urgent need to simplify the procedures and synthesis of mixed oxide pillars containing Ni atoms.

Based on an understanding of the stability and synthesis of the PILC process, functional AlNi-PILC supports were prepared using a high temperature and high pressure hydrothermal method. Compared with MMT, AlNi-PILC displayed a larger specific surface area, a larger pore volume, and a high thermal stability. Therefore, it could be used as support to prepare the high performance catalyst. The influence of the introduction of CeO_2 into Pd/AlNi-PILC for benzene combustion was also studied. The relationship between texture-structure and catalytic properties was systematically characterized and analyzed by X-ray diffraction (XRD), N_2 adsorption/desorption, high resolution transmission electron microscopy and energy dispersive X-ray spectroscopy (HRTEM-EDS), the temperature-programmed desorption of benzene (benzene-TPD), and the in-situ temperature-programmed surface reaction of benzene (benzene-TPSR) experiments. The water and chlorobenzene were systematically studied in order to preliminarily explore the Pd/Ce/AlNi-PILC potential for further industrial application.

2. Experimental

2.1. Synthesis

MMT was used as initial material and the AlNi-pillaring agent was prepared using a hydrothermal method. The aqueous solution of $Ni(NO_3)_2 \cdot 6H_2O$ and Locron L from Clariant (containing 6 mol/L Al ions) was mixed in autoclave (the molar ratio was Al/Ni = 5:1), and deionized water was added so that the concentration of Al ion was 2.0 mol/L. The autoclave was placed in an oven at 100 °C for 16 h and subsequently cooled down to 30 °C. The maintained solution was diluted to 600 mL and, finally, AlNi-pillaring agent was obtained. The following preparation of AlNi-PILC by the similar method was detailed in our previous research [6]. The $X\%$ Ce/AlNi-PILC samples were prepared by impregnation of $Ce(NO_3)_2 \cdot 6H_2O$ (X = 2.5, 5, 7.5, 10, 12.5, and 15, respectively). After keeping impregnated samples at 30 °C for 12 h, the samples were dried at 110 °C and subsequently calcined at 400 °C for 2 h. The Pd/$X\%$ Ce/AlNi-PILC samples were obtained by impregnating $X\%$ Ce/AlNi-PILC with an aqueous H_2PdCl_4 solution at 30 °C for 12 h, and the yellow was completely disappeared under an infrared lamp. Then, 5% hydrazine hydrate was added and reacted for 3 h, and the samples were filtered and washed by deionized water until no Cl^- was detected in the filtrate by aqueous $AgNO_3$ solution. Samples were dried at 110 °C, and subsequently calcined at 400 °C for 2 h. The Pd content of all catalysts was 0.2 wt. %.

2.2. Catalytic Activity Tests

The experiments were performed with a 350 mg catalyst in a WFS-3010 microreactor (Xianquan, Tianjin, China). An analysis of the reactants and products was performed by on line gas chromatography (Shimadzu, GC-14C, Kyoto, Japan) with a flame ionization detector (FID). The reactive flow (120 mL/min) was composed of gaseous benzene (1000 ppm) in dry air with a gas hourly space velocity (GHSV) of 20,000 h^{-1}. The data were recorded and analyzed using an N2000 chromatography data workstation. The catalytic activity was determined by parallel analytical measurement at a certain temperature (parallel determination of three identical catalysts, approximately 0.5 h per series), and the average was taken as the final conversion. And the benzene conversion was calculated as follows: benzene conversion (%) = $\frac{[\text{benzene}]\text{in} - [\text{benzene}]\text{out}}{[\text{benzene}]\text{in}} \times 100\%$ (where $[\text{benzene}]_{\text{in}}$ is the benzene concentration in the feed gas, and $[\text{benzene}]_{\text{out}}$ is the benzene concentration in the products).

In order to study the "mixture effect" of the feed gas, 100 ppm chlorobenzene and 10,000 ppm water vapor were introduced, respectively. The any possible combustion products were further detected by mass spectrometry (MS, QGA, Hiden, Warrington, UK). H2O and CO_2 were the only detected byproducts, and thus conversion was calculated based on benzene consumption. The durability of catalysts for benzene combustion was also investigated under the same condition.

2.3. Characterization

The samples were characterized by the XRD technique (PANalytical, Almelo, The Netherlands) for the d_{001} value and phase composition. The specific surface area (S_{BET}), mesoporous area (A_{mes}), total pore volume (V_{p}), micropore volume (V_{mic}), and pore size distribution of the samples were determined by N_2 adsorption isotherms. High-resolution transmission electron microscopy (HRTEM, JEOL, Valley, Japan) was employed to get the catalyst morphology and particle size. The chemical compositions of the catalysts were determined with energy dispersive X-ray spectroscopy (EDS, JEOL, Valley, Japan). All the characterization methods for the samples have been reported and detailed in our previous research [3,9,10]. The palladium and Ce contents were measured by an Inductively Coupled Plasma Optical Emission Spectrometer (ICP-OES, Leeman Labs, Hudson, NH, USA) after the dissolution of the catalysts in a mixture of HF and HNO_3 solution. Benzene-TPD and the benzene-TPSR) experiments were performed in a quartz tube. Prior to adsorption of benzene, the catalyst (350 mg) was pretreated in dry air at 300 °C for 0.5 h. After being cooled down to 50 °C, the adsorption of benzene was carried out under a flow of N_2/benzene (TPD) or ($20\%O_2$/Ar) /benzene (TPSR) until adsorption saturation, as indicated by the stable signal of benzene in the mass spectrometer. Then, a pure N_2 flow was carried out for 1h to clean the benzene in the pipe. Finally, the desorption or oxidation of benzene was implemented followed by a flow of pure N_2 (TPD) or ($20\%O_2$/Ar)/benzene (TPSR) by a step of 7.5 °C/min from 50 to 500 °C. The concentration of benzene and the final products (CO_x and H_2O) were measured on-line by MS.

3. Results and Discussion

3.1. Catalytic Performance and Stability Test

Generally, benzene is completely degraded at 600 °C under a no catalysts condition. The catalytic activity of catalysts for benzene combustion is displayed in Figure 1a. It can be seen that Pd/MMT exhibits poor performance and the complete conversion of benzene does not occur until 350 °C. Pd/AlNi-PILC is able to completely degrade benzene at 318 °C. The results suggest that AlNi-PILC is more suitable to be a catalytic support. In addition, Ce doping significantly improved the catalytic activities of Pd/MMT and Pd/AlNi-PILC. Therefore, the effect of Ce content was also investigated in Figure 1b. According to the values of $T_{98\%}$ (temperature which benzene conversion reaches 98%), the order for the catalytic activity is Pd/AlNi-PILC (318 °C) < Pd/2.5% Ce/AlNi-PILC (310 °C) < Pd/5% Ce/AlNi-PILC (290 °C) < Pd/7.5% Ce/AlNi-PILC (270 °C) < Pd/10% Ce/AlNi-PILC (260 °C) < Pd/15% Ce/AlNi-PILC (255 °C) < Pd/12.5% Ce/AlNi-PILC (240 °C). The results further demonstrate that the addition of various amounts of Ce improves the catalytic activities of the Pd/AlNi-PILC catalysts. When Ce loading is <12.5%, the activity of the catalyst increases with the addition of Ce content. However, the catalytic activity decreases when Ce loading is 15%, which is due to the fact that Ce acts as a promoting component and too much Ce loading may override the PdO active sites.

Similar results have been reported in our previous research [40–42]. Thus, proper Ce content and the metal-metal interaction can significantly enhance the Pd/Ce/AlNi-PILC activity. Particularly, the catalyst of 12.5% Ce loading exhibits the highest activity and the $T_{98\%}$ of benzene conversion is 240 °C. In addition, the catalytic combustion performances of benzene over some typical noble metal-based catalysts [12,15,43–45] were listed in Table S1, and Pd/12.5% Ce/AlNi-PILC catalyst possesses better performance.

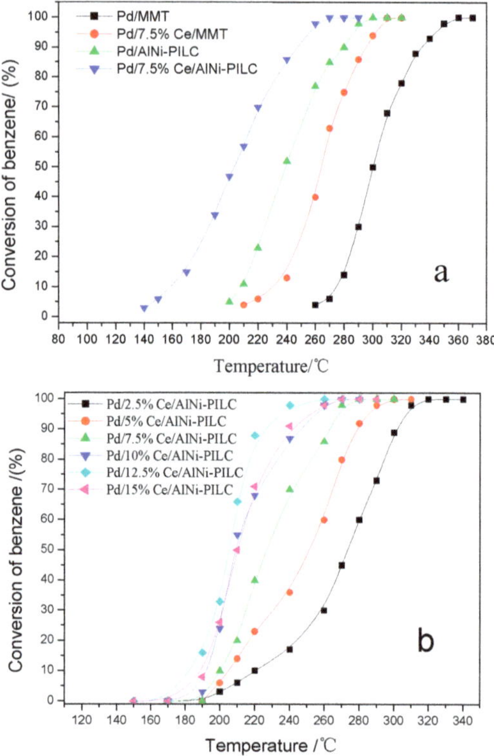

Figure 1. (**a**) Influence of the addition of Ce into Pd/MMT and Pd/AlNi-PILC for benzene complete oxidation; (**b**) influence of the content of Ce into Pd/AlNi-PILC for benzene complete oxidation. Benzene concentration: 1000 ppm; gas hourly space velocity (GHSV): 20,000 h^{-1}; Catalyst amount: 350 mg.

Figure 2 presents the lifetime test result for the most active catalyst (Pd/12.5% Ce/AlNi-PILC) at 230 °C for 1000 h. The conversion of benzene remained at around 94% and no obvious deactivation was observed, which demonstrates that it exhibits stable catalytic activity for combustion. Moreover, in practice, water and chlorinated VOCs always exist in the waste gases. As shown in Figure 2, in the first continuous 100 h reaction in the presence of 100 ppm chlorobenzene, the activity of the catalyst decreases slightly, because of the effect of competitive adsorption and oxidation on the active sites. When 10,000 ppm water is introduced, its activity further decreases, due to the competitive adsorption effect. Interestingly, the catalytic activity is recovered to the initial level after water and chlorobenzene are removed. In all, these results mentioned above confirm that Pd/12.5% Ce/AlNi-PILC presents a wide range of possibilities for further industrial application, due to its high catalytic performances for combustion of both non-chlorinated VOCs and chlorinated VOCs, as well as its good resistance to water.

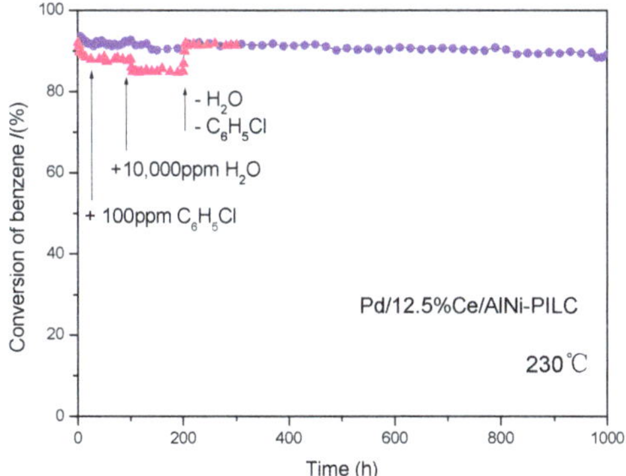

Figure 2. Lifetime test performed for Pd/12.5% Ce/AlNi-PILC at 230 °C.

3.2. Effects of Loading Pd Content and Benzene Concentration

Figure S1 presents the effects of Pd loading contents on the benzene catalytic performance of X% Pd/12.5% Ce/AlNi-PILC. Benzene conversion increases with the increase of Pd content from 0.2 to 0.5%. When the content of Pd increases to 0.5%, the activity changes slightly compared to 0.4% Pd catalyst. It suggests that the addition of proper Pd content can present high dispersion on support and improve the catalytic performances obviously. Figure S2 presents the effects of benzene inlet concentrations (500 to 2500 ppm) on the catalytic performance of 0.2% Pd/12.5% Ce/AlNi-PILC. Benzene conversion increases appreciably with the increase of its inlet concentrations from 500 to 1500 ppm. When low concentration benzene is fed, the amount of chemisorbed benzene on catalyst active sites is low and can be a reaction controlling factor. However, chemisorbed oxygen on the catalyst active sites can become the reaction controlling factor when the benzene concentration increases to a certain point, thus the conversion of benzene should be prohibited.

3.3. XRD and N_2 Adsorption Analysis

The small-angle XRD patterns are obtained to the confirm two-dimensional layered structure of MMT and AlNi-PILC in Figure 3. It can be seen that the d_{001} spacing of MMT and AlNi-PILC is 1.26 nm and 2.11 nm, respectively, with corresponding 2θ values of 6.99° and 4.18°. It suggests that AlNi poly-cations intercalated between the layers are much larger than Na ions in size, leading to a layered structure with large spacing. Thus, the synthesis of pillared clays is successful.

Figure 4 shows the high-angle XRD patterns of samples. Both MMT and AlNi-PILC contain similar cristobalite peaks (23.3°) and quartz peaks (26.5°), which are the characteristic peaks of montmorillonite. It suggests that pore structure is well preserved during the synthesis of AlNi-PILC. It must also be mentioned that a Al-Si polymeride peak (28.14°) disappears when $NiAl_2O_4$ peaks are formed by the interaction between nickel oxide and alumina during the calcination process. These peaks appear at 37.0°, 59.7°, and 64.5°, respectively, and most were overlapped with the peaks of Al_2O_3 [46]. Thus, some feature diffraction peaks of Al_2O_3 became weaker. Studies [46] have shown that spinal $NiAl_2O_4$ also promotes Pd catalysts to be more resistant to deactivation, which has been proved in the above lifetime test (Figure 2). In addition, neither palladium nor palladium oxide is detected for catalysts containing Pd. The result shows that the PdO particles were small, which is below the X-ray detection range, or Pd was highly dispersed on the supports. Notably, for the Pd/12.5% Ce/AlNi-PILC catalyst, the intensity of CeO_2 peaks becomes significantly weaker than Pd/12.5% Ce/MMT, revealing higher dispersion of CeO_2 on AlNi-PILC. From the catalytic activity and XRD results, it can be observed that better dispersion of CeO_2 on AlNi-PILC is one of the key factors to enhance the catalytic activity of the catalysts. The above results also indicate that AlNi-PILC is useful for the high dispersion of PdO and CeO_2 on its surface, which is further confirmed by below HRTEM images.

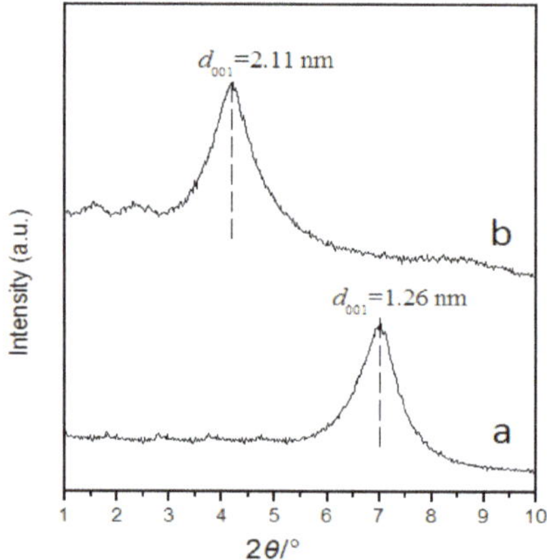

Figure 3. Small-angle X-ray diffraction (XRD) patterns. (**a**) Montmorillonite (MMT); (**b**) AlNi-PILC.

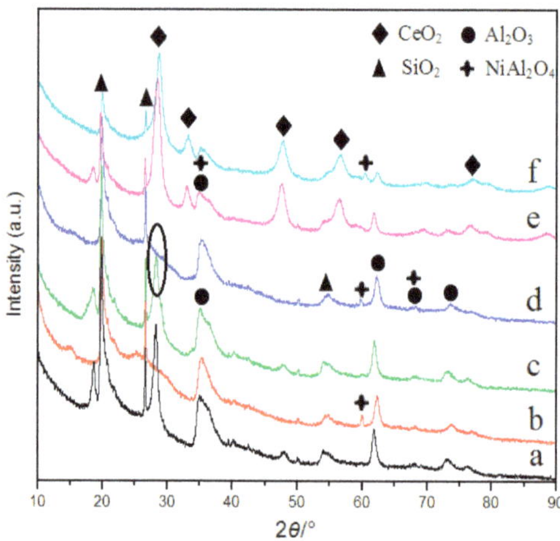

Figure 4. Long-angle XRD patterns. (**a**) MMT; (**b**) AlNi-PILC; (**c**) Pd/MMT; (**d**) Pd/AlNi-PILC; (**e**) Pd/12.5% Ce/MMT; (**f**) Pd/12.5% Ce/AlNi-PILC.

The textural properties of all samples were characterized by the N_2 adsorption/desorption method (Figure 5). Textural properties are listed in Table 1, and it can be seen that all samples provided the IV type isotherm of the miro-mesoporous materials with a sharp ramp in the relative pressure of 0.45, which is due to the capillary condensation of nitrogen in pores [47]. Additionally, S_{BET}, A_{mes}, and V_{mic} of all samples are summarized in Table 1. The S_{BET} of AlNi-PILC reaches to 374.8 m^2/g, which is much larger than MMT. In addition, the S_{BET} of two synthesized samples (Pd/AlNi-PILC and Pd/12.5% Ce/AlNi-PILC) decreases to 298.4 m^2/g and 230.5 m^2/g, respectively, in which doped cations may enter the micro-mesopores of AlNi-PILC. Some similar results about the incorporation of other metals (such as Co, Nd, La, etc.) into the silicate framework are reported [48,49]. It is important to highlight that the values of S_{BET} are not connected with the order for activity of benzene combustion, revealing that it is not the only factor affecting catalytic oxidative performance.

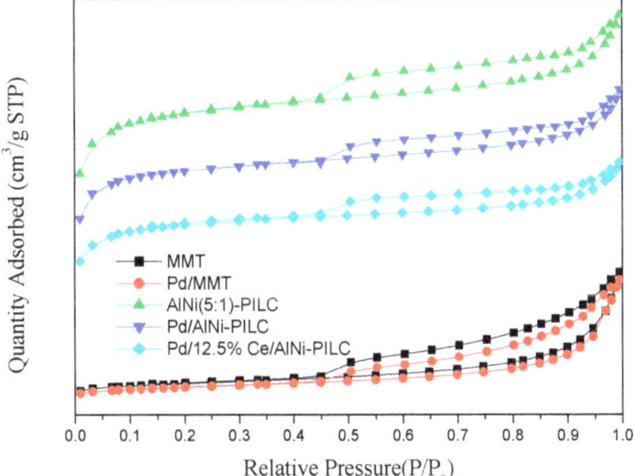

Figure 5. N_2 adsorption/desorption isotherms of the samples.

Table 1. Characteristics of the samples: surface area and pore volume.

Samples	S_{BET} [a] (m^2/g)	A_{mes} [b] (m^2/g)	V_p [c] (cm^3/g)	V_{mic} [d] (cm^3/g)
MMT	24.6	20.1	0.054	0.0018
AlNi-PILC	374.8	86.1	0.212	0.1336
Pd/MMT	20	17.9	0.05	0.0007
Pd/AlNi-PILC	298.4	61.4	0.17	0.1101
Pd/12.5% Ce/AlNi-PILC	230.5	56.8	0.131	0.0804

[a] BET specific surface area; [b] Calculated from BJH method; [c] Total pore volume estimated at $P/P_0 = 0.99$; [d] Calculated from the t-method.

Table 1 also shows that all samples possess different V_{mic} of 0.0018–0.1336 cm^3/g and the V_p of 0.054 to 0.212 cm^3/g, respectively. The large S_{BET} (374.8 m^2/g) and V_P (0.212 cm^3/g) of the AlNi-PILC are responsible for the well dispersion of the metallic particles and exposing more metallic sites on the surface for catalytic application. This is supported by the higher catalytic activity of the Pd/12.5% Ce/AlNi-PILC catalyst (Figure 1 and Table 1). In conclusion, the textural characterization results suggest that some Pd^{2+} and Ce^{4+} reach the inner porous network of AlNi-PILC, resulting in a strong interaction among PdO, CeO_2, and AlNi-PILC.

3.4. HRTEM Analysis

Figure 6 presents HRTEM pictures and the EDS spectra of MMT, Pd/MMT, Pd/AlNi-PILC, and Pd/12.5% Ce/AlNi-PILC. The HRTEM image of AlNi-PILC shows that the material has a better-ordered hexagonal arrays structure than MMT. HRTEM and the Map data picture of Pd/MMT show that PdO particles have more serious aggregation than Pd/AlNi-PILC. Also, the incorporation of CeO_2 to the Pd catalyst leads to a higher PdO dispersion than that of Pd/AlNi-PILC, which is the result of the interaction with CeO_2 and PdO. For the Pd/12.5% Ce/AlNi-PILC catalyst, in spite of its high Ce concentration, well dispersed PdO and CeO_2 nanoparticles are obtained, which is consistent with the XRD result. In addition, after the addition of Ce and Pd, the particle sizes of AlNi-PILC do not change significantly and still present in an ordered framework. The Map data image of Pd/12.5% Ce/AlNi-PILC clearly reveals the well-dispersed active ingredients on AlNi-PILC. Pd and Ce were identified in the EDS spectra, confirming the successful loading of the active ingredients on the surface of AlNi-PILC. On the basis of the above results, we conclude that a higher dispersion of $PdO-CeO_2$ on AlNi-PILC may be responsible for the good activity of these samples.

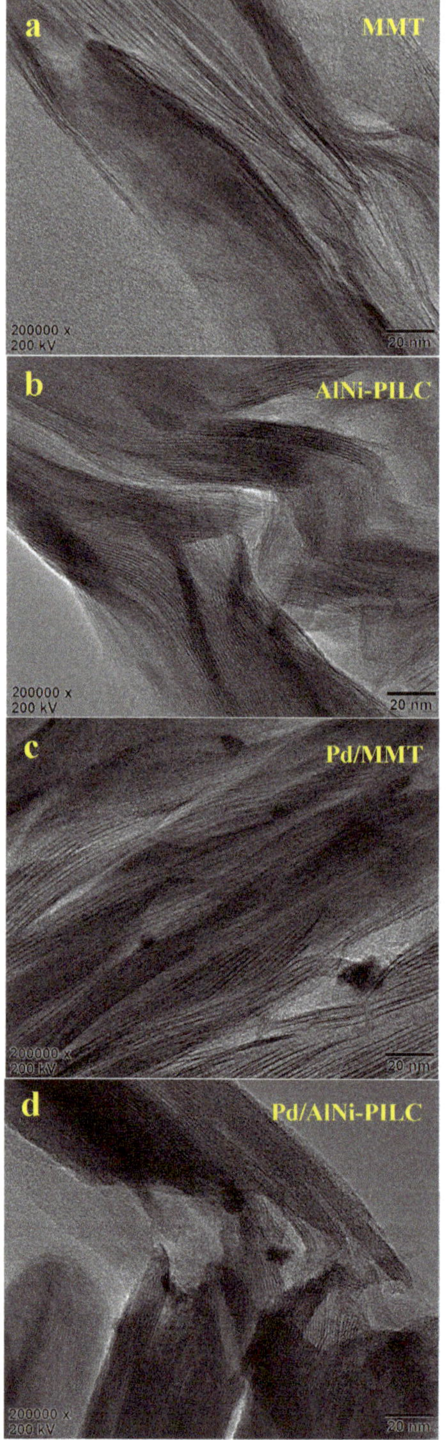

Figure 6. *Cont.*

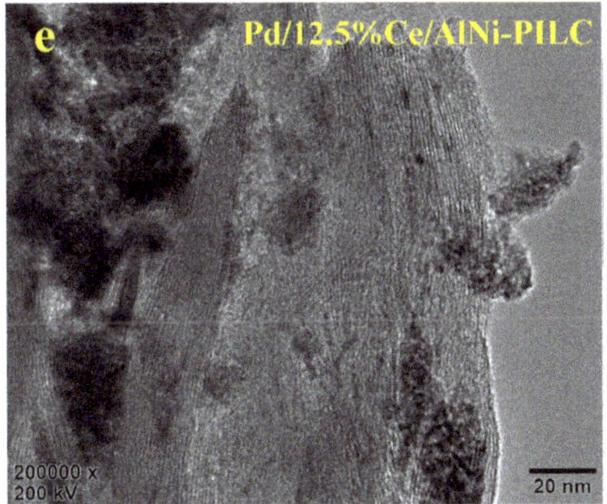
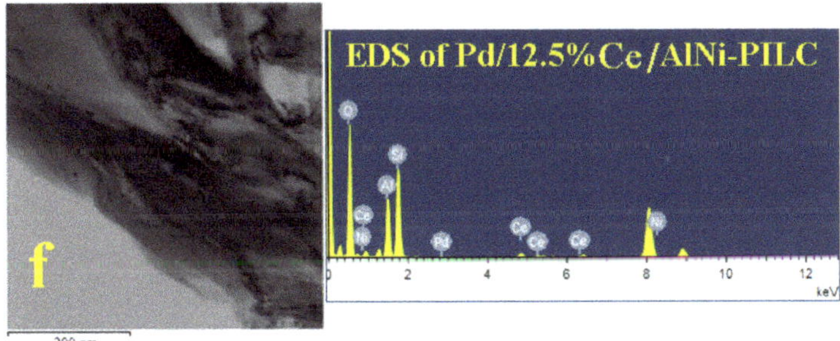

Figure 6. High resolution transmission electron microscopy (HRTEM) pictures. (**a**) MMT; (**b**) AlNi-PILC; (**c**) Pd/MMT; (**d**) Pd/AlNi-PILC; (**e**) Pd/12.5% Ce/AlNi-PILC; (**f**) the energy dispersive X-ray spectroscopy (EDS) spectra.

3.5. ICP-OES Analysis

The nominal mass percentage of Pd for all the catalysts is 0.2 and the Ce for Pd/12.5% Ce/AlNi-PILC is 12.5%, respectively. The real metal loadings of different catalysts measured by ICP-OES are listed in Table S2. The real concentration of Pd in all the catalysts is about 0.19%. In the fresh Pd/12.5% Ce/AlNi-PILC catalyst, the real contents of Pd and Ce are 0.187% and 12.0%, respectively, and no significant variation is observed after reaction. This indicates that AlNi-PILC can stabilize the active phase, and that this catalyst is reusable in benzene combustion and can perform well in a wide range of applications for the combustion of VOC.

3.6. TPD Analysis

Figure 7a shows the adsorption of benzene ($m/z = 78$) profiles of the catalysts in the TPD test. Compared with the MMT catalyst, the amount of benzene adsorption increases sharply when AlNi-PILC is used as support, and the order is Pd/AlNi-PILC > Pd/12.5% Ce/AlNi-PILC > Pd/MMT. This is probably because the large interlayer distance and pore volume of AlNi-PILC is advantageous for the adsorption of benzene. By integrating over the adsorption peaks, the benzene adsorption capacities of Pd/MMT, Pd/AlNi-PILC, and Pd/12.5% Ce/AlNi PILC are calculated to be about 11.0, 28.5, and 36.8 µmol/g, respectively. Compared to the Pd/AlNi-PILC catalyst, the amount of benzene adsorption is decreased over Pd/12.5% Ce/AlNi-PILC, which may be caused by the decrease in specific surface area and the pore volume. Figure 7b presents the Pd/12.5% Ce/AlNi-PILC that exhibits a high desorption temperature, which leads to a high adsorption strength and catalytic activity. In addition, the desorption temperature for the reactant over the catalysts should have a great influence on the catalytic activity of the catalysts. Generally, the closer the desorption temperatures of benzene and O_2 to the combustion temperature of benzene, the higher benzene conversion will be achieved. The conclusion will be further investigated, as shown below.

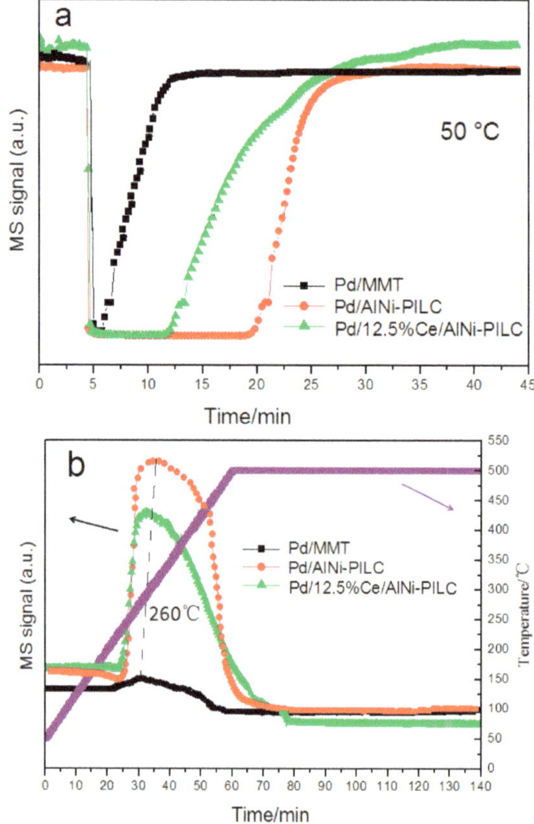

Figure 7. (**a**) Benzene (m/z = 78) adsorption profiles in temperature-programmed desorption (TPD) over Pd/AlNi-PILC, Pd/12.5% Ce-AlNi-PILC, and Pd/MMT catalysts; (**b**) benzene (m/z = 78) desorption profiles in TPD over Pd/AlNi-PILC, Pd/12.5% Ce-AlNi-PILC, and Pd/MMT catalysts.

3.7. Benzene-TPSR Analysis

As is well known, the catalytic process is a dynamic and in-situ surface reaction. Thus, in order to investigate the oxidative performances of the catalysts under the dynamic condition and get more information on the real oxidation process, as well as the adsorption/desorption and oxidizing properties of the catalysts for benzene combustion, the evolution of any possible organic byproducts and the final products (CO_x and H_2O) on the catalyst surface are evaluated by the in-situ TPSR technique [11].

As shown in Figure 8, in the range of 50 °C to 200 °C, the signal of benzene in the Pd/12.5% Ce/AlNi-PILC catalyst is observed while the signal of CO_2 is absent, indicating that only the desorption of benzene occurs. Compared with the TPD results, the peak temperature of benzene over three catalysts is shifted to a lower temperature in the presence of gas phase O_2, and the temperature for desorption of the benzene signal decreases in the order of Pd/12.5% Ce/AlNi-PILC $\approx$ Pd/AlNi-PILC

> Pd/MMT. The results reveal that the strong interaction between AlNi-PILC and metals enhances benzene adsorption.

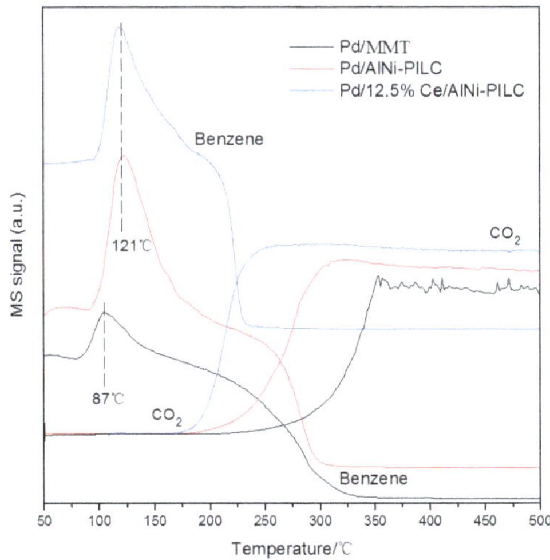

Figure 8. Results of benzene-TPSR characterization for benzene combustion over the catalysts.

In addition, the desorption peak of benzene over Pd/MMT is smaller than Pd/12.5% Ce/AlNi-PILC, implying that its adsorption capacity is lower, which is not beneficial for the benzene oxidation reaction. As the reaction temperature increases, the oxidation of benzene gradually becomes obvious, due to the detection of CO_2. The final products (CO_x and H_2O) were measured on–line by MS and the result indicates that CO_2 is the only carbon product. It indicates that the above catalysts have high selectivity and high activity. Moreover, the temperature for the disappearance of the benzene signal increase is in the order of Pd/12.5% Ce/AlNi-PILC (240 °C) < Pd/AlNi-PILC (315 °C) < Pd/MMT (350 °C), and the temperature for the appearance of the CO_2 signal increase is consistent with the above order. It is noteworthy that for Pd/12.5% Ce/AlNi-PILC, the degradation temperature (240 °C) of benzene is lower than the desorption temperature (260 °C), which helps complete benzene combustion during the desorption process, so it exhibits high catalytic activity.

4. Conclusions

In this work, AlNi-PILC material and Pd/Ce/AlNi-PILC catalysts with different Ce content were successfully synthesized and used in the catalytic combustion of low concentration benzene. The structure and redox properties of these materials were characterized by XRD, N_2 adsorption, HRTEM-EDS, TPD, and TPSR techniques. XRD and N_2 adsorption results indicate that AlNi-PILC material shows higher ordered hexagonal pore structure and higher S_{BET} than MMT. Also, Pd-Ce-supported catalysts still maintain ordered layer structures. From the HRTEM-EDS results, the incorporation of CeO_2 to the Pd catalysts leads to a higher dispersion than that of Pd/AlNi-PILC. The appropriate crystallized size of AlNi-PILC support and the high dispersed PdO nanosize particles might have a large significance for Pd/12.5% Ce/AlNi-PILC stability and catalytic activity. The TPD and TPSR results show that the Pd/12.5% Ce/AlNi-PILC high capacity for adsorption/desorption-catalytic combustion of benzene are due to the high benzene adsorption strength and the similar temperature between benzene desorption and the combustion process. Therefore, Pd/12.5% Ce/AlNi-PILC can complete benzene combustion at 240 °C. Furthermore, stability tests indicate that there is no obvious deactivation for the Pd/12.5% Ce/AlNi-PILC catalyst in the

1000 h continuous reaction, whether in the water condition or in the presence of C_7H_8, which indicates that it deserves more attention and that there is potential for industrial application.

Supplementary Materials: The following are available online at www.mdpi.com/2072-6651/9/8/949/s1, Figure S1: Effects of Pd content on catalytic activity of Pd/12.5% Ce/AlNi-PILC for benzene combustion. Benzene concentration: 1000 ppm; GHSV: 20,000 h^{-1}; Catalyst amount: 350 mg. Figure S2: Effects of inlet concentration on benzene catalytic combustion over Pd/12.5% Ce/AlNi-PILC. Benzene concentration: 500–2500 ppm; GHSV: 20,000 h^{-1}; Catalyst amount: 350 mg. Table S1: Main data of reported literatures on catalytic combustion of benzene over supported noble metal catalysts. Table S2: Metal loadings (wt. %) of different catalysts.

Acknowledgments: The authors would like to thank the support provided by National Natural Science Foundation of China (Project 21577094).

Author Contributions: Shufeng Zuo designed the experiments; Jingrong Li and Peng Yang performed the experiments; Jingrong Li and Shufeng Zuo wrote the paper; Chenze Qi provided some experimental equipment and guided the experiments.

Conflicts of Interest: The authors declare no conflict of interest.

References

1. Lindgren, T. A case of indoor air pollution of ammonia emitted from concrete in a newly built office in Beijing. *Build. Environ.* **2010**, *45*, 596–600. [CrossRef]

2. Zhang, Z.X.; Jiang, Z.; Shang guan, W.F. Low-temperature catalysis for VOCs removal in technology and application: A state-of-the-art review. *Catal. Today* **2016**, *264*, 270–278. [CrossRef]

3. Yang, P.; Yang, S.S.; Shi, Z.N.; Tao, F.; Guo, X.L.; Zhou, R.X. Accelerating effect of ZrO_2 doping on catalytic performance and thermal stability of CeO_2–CrO_x mixed oxide for 1,2-dichloroethane elimination. *Chem. Eng. J.* **2016**, *285*, 544–553. [CrossRef]

4. Aznárez, A.; Korili, S.A.; Gil, A. The promoting effect of cerium on the characteristics and catalytic performance of palladium supported on alumina pillared clays for the combustion of propene. *Appl. Catal. A Gen.* **2014**, *474*, 95–99. [CrossRef]

5. Castaño, M.H.; Molina, R.; Moreno, S. Cooperative effect of the Co-Mn mixed oxides for the catalytic oxidation of VOCs: Influence of the synthesis method. *Appl. Catal. A Gen.* **2015**, *492*, 48–59. [CrossRef]

6. Ding, M.L.; Zuo, S.F.; Qi, C.Z. Preparation and characterization of novel composite AlCr-pillared clays and preliminary investigation for benzene adsorption. *Appl. Clay Sci.* **2015**, *115*, 9–16. [CrossRef]

7. Yang, P.; Xue, X.M.; Meng, Z.H.; Zhou, R.X. Enhanced catalytic activity and stability of Ce doping on Cr supported HZSM-5 catalysts for deep oxidation of chlorinated volatile organic compounds. *Chem. Eng. J.* **2013**, *234*, 203–210. [CrossRef]

8. Yang, H.G.; Deng, J.G.; Xie, S.H.; Jiang, Y.; Dai, H.X.; Au, C.T. Au/MnO$_x$/3DOM SiO$_2$: Highly active catalysts for toluene oxidation. *Appl. Catal. A Gen.* **2015**, *507*, 139–148. [CrossRef]

9. Zuo, S.F.; Du, Y.J.; Liu, F.J.; Han, D.; Qi, C.Z. Influence of ceria promoter on shell-powder-supported Pd catalyst for the complete oxidation of benzene. *Appl. Catal. A Gen.* **2013**, *451*, 65–70. [CrossRef]

10. Yang, P.; Zuo, S.F.; Shi, Z.N.; Tao, F.; Zhou, R.X. Elimination of 1,2-dichloroethane over (Ce, Cr)$_x$O$_2$/MO$_y$ catalysts (M = Ti, V, Nb, Mo, W and La). *Appl. Catal. B Environ.* **2016**, *191*, 53–61. [CrossRef]

11. Yang, P.; Yang, S.S.; Shi, Z.N.; Meng, Z.H.; Zhou, R.X. Deep oxidation of chlorinated VOCs over CeO_2-based transition metal mixed oxide catalysts. *Appl. Catal. B Environ.* **2015**, *162*, 227–235. [CrossRef]

12. Zhao, S.; Li, K.; Jiang, S.; Li, J. Pd-Co based spinal oxides derived from Pd nanoparticles immobilized on layered double hydroxides for toluene combustion. *Appl. Catal. B Environ.* **2016**, *181*, 236–248. [CrossRef]

13. Li, P.; He, C.; Cheng, J.; Ma, C.Y.; Dou, B.J.; Hao, Z.P. Catalytic oxidation of toluene over Pd/Co$_3$AlO catalysts derived from hydrotalcite-like compounds: Effects of preparation methods. *Appl. Catal. B Environ.* **2011**, *101*, 570–579. [CrossRef]

14. Deng, J.; He, S.; Xie, S.; Yang, H.; Liu, Y.; Guo, G.; Dai, H. Ultralow Loading of Silver Nanoparticles on Mn_2O_3 Nanowires Derived with Molten Salts: A High-Efficiency Catalyst for the Oxidative Removal of Toluene. *Environ. Sci. Technol.* **2015**, *49*, 11089–11095. [CrossRef] [PubMed]

15. Liu, F.J.; Zuo, S.F.; Wang, C.; Li, J.T.; Xiao, F.S.; Qi, C.Z. Pd/transition metal oxides functionalized ZSM-5 single crystals with *b*-axis aligned mesopores: Efficient and long-lived catalysts for benzene combustion. *Appl. Catal. B Environ.* **2014**, *148–149*, 106–113. [CrossRef]

16. Xing, T.; Wan, H.Q.; Shao, Y.; Han, Y.X.; Xu, Z.Y.; Zheng, S.R. Catalytic combustion of benzene over γ-alumina supported chromium oxide catalysts. *Appl. Catal. A Gen.* **2013**, *468*, 269–275. [CrossRef]

17. Tsoncheva, T.; Issa, G.; Blasco, T.; Dimitrov, M.; Popova, M.; Hernández, S.; Kovacheva, D.; Atanasova, G.; Nieto, J.L. Catalytic VOCs elimination over copper and cerium oxide modified mesoporous SBA-15 silica. *Appl. Catal. A Gen.* **2013**, *453*, 1–12. [CrossRef]

18. Kim, S.C.; Shim, W.G. Catalytic combustion of VOCs over a series of manganese oxide catalysts. *Appl. Catal. B Environ.* **2010**, *98*, 180–185. [CrossRef]

19. Shokouhimehr, M. Magnetically Separable and Sustainable Nanostructured Catalysts for Heterogeneous Reduction of Nitroaromatics. *Catalysts* **2015**, *5*, 534–560. [CrossRef]

20. Mahdi, R.S.; Aejung, A.; Mohammadreza, S. Gadolinium Triflate Immobilized on Magnetic Nanocomposites as Recyclable Lewis Acid Catalyst for Acetylation of Phenols. *Nanosci. Nanotechnol. Lett.* **2014**, *6*, 309–313.

21. Kim, A.; Rafiaei, S.M.; Abolhosseini, S.; Shokouhimehr, M. Palladium Nanocatalysts Confined in Mesoporous Silica for Heterogeneous Reduction of Nitroaromatics. *Energy Environ. Focus* **2015**, *4*, 18–23. [CrossRef]

22. Shokouhimehr, M.; Shin, K.Y.; Lee, J.S.; Hackett, M.J.; Jun, S.W.; Oh, M.H.; Jang, J.; Hyeon, T. Magnetically recyclable core—Shell nanocatalysts for efficient heterogeneous oxidation of alcohols. *J. Mater. Chem. A* **2014**, *2*, 7593–7599. [CrossRef]

23. Zhao, W.; Cheng, J.; Wang, L.; Chu, J.L.; Qu, J.K.; Liu, Y.H.; Li, S.H.; Zhang, H.; Wang, J.C.; Hao, Z.P.; et al. Catalytic combustion of chlorobenzene on the Ln modified Co/HMS. *Appl. Catal. B Environ.* **2012**, *127*, 246–254. [CrossRef]

24. Sedjame, H.J.; Fontaine, C.; Lafaye, G.; Barbier, J., Jr. On the promoting effect of the addition of ceria to platinum based alumina catalysts for VOCs oxidation. *Appl. Catal. B Environ.* **2014**, *144*, 233–242. [CrossRef]

25. Ferreira, D.; Zanchet, R.; Rinaldi, U.; Schuchardt, S. Effect of the CeO_2 content on the surface and structural properties of CeO_2–Al_2O_3 mixed oxides prepared by sol gel method. *Appl. Catal. A Gen.* **2010**, *388*, 45–56. [CrossRef]

26. Shyu, J.Z.; Otto, K. Characterization of Pt/γ-alumina catalysts containing ceria. *J. Catal.* **1989**, *115*, 16–23. [CrossRef]

27. Yermakov, Y.I.; Kuznetsov, B.N. Supported metallic catalysts prepared by decomposition of surface ornagometallic complexes. *J. Mol. Catal.* **1980**, *9*, 13–40. [CrossRef]

28. Ballini, R.; Fiorini, D.; Victoria, G.M.; Palmier, A. Michael addition of α-nitro ketones to conjugated enones under solventless conditions using silica. *Green Chem.* **2003**, *5*, 475–476. [CrossRef]

29. Rani, R.V.; Srinivas, N.; Kisham, R.M.; Kulkarni, S.J.; Raghavan, K.V. Zeolite-catalyzed cyclocondensation reaction for the selective synthesis of 3,4-dihydropyrimidin-2(1H)-ones. *Green Chem.* **2001**, *3*, 305–306. [CrossRef]

30. Choudhary, V.R.; Patil, K.Y.; Jana, S.K. Acylation of aromatic alcohols and phenols over $InCl_3$/montmorillonite K-10 catalysts. *J. Chem. Sci.* **2004**, *116*, 175–177. [CrossRef]

31. Tong, D.S.; Zhou, C.H.; Lu, Y.; Yu, H.; Zhang, G.F.; Yu, W.H. Adsorption of Acid Red G dye on octadecyl trimethylammonium montmorillonite. *Appl. Clay Sci.* **2010**, *50*, 427–431. [CrossRef]

32. Molina, C.B.; Casas, J.A.; Zazo, J.A.; Rodriguez, J.J. A comparison of Al-Fe and Zr-Fe pillared clays for catalytic wet peroxide oxidation. *Chem. Eng. J.* **2006**, *118*, 29–35. [CrossRef]

33. Babu, B.H.; Lee, M.; Hwang, D.W.; Kim, Y.; Chae, H.J. An integrated process for production of jet fuel range olefins from ethylene using Ni-Al-SBA-15 and Amberlyst-35 catalysts. *Appl. Catal. A Gen.* **2017**, *530*, 48–55. [CrossRef]

34. Tanaka, M.; Itadani, A.; Kuroda, Y.; Iwamoto, M. Effect of Pore Size and Nickel Content of Ni-MCM-41 on Catalytic Activity for Ethene Dimerization and Local Structures of Nickel Ions. *J. Phys. Chem. C* **2012**, *116*, 5664–5672. [CrossRef]

35. Lallemand, M.; Rusu, O.A.; Dumitriu, E.; Finiels, A.; Fajula, F.; Hulea, V. NiMCM-36 and NiMCM-22 catalysts for the ethylene oligomerization: Effect of zeolite texture and nickel cations/acid sites ratio. *Appl. Catal. A Gen.* **2008**, *338*, 37–43. [CrossRef]

36. Martínez, A.; Arribas, M.A.; Concepción, P.; Moussa, S. New bifunctional Ni-H-Beta catalysts for the heterogeneous oligomerization of ethylene. *Appl. Catal. A Gen.* **2013**, *467*, 509–518. [CrossRef]

37. Lin, S.; Shi, L.; Zhang, H.; Zhang, N.; Yi, X.; Zheng, A.; Li, X. Tuning the pore structure of plug-containing Al-SBA-15 by post-treatment and its selectivity for C_{16} olefin in ethylene oligomerization. *Microporous Mesoporous Mater.* **2014**, *184*, 151–161. [CrossRef]

38. Andrei, R.D.; Popa, M.I.; Fajula, F.; Hulea, V. Heterogeneous oligomerization of ethylene over highly active and stable Ni-Al-SBA-15 mesoporous catalysts. *J. Catal.* **2015**, *323*, 76–84. [CrossRef]

39. Aznarez, A.; Delaigle, R.; Eloy, P.; Gaigneaux, E.M.; Korili, S.A.; Gil, A. Catalysts based on pillared clays for the oxidation of chlorobenzene. *Catal. Today* **2015**, *246*, 15–27. [CrossRef]

40. Zuo, S.F.; Zhou, R.X. Influence of synthesis condition on pore structure of Al pillared clays and supported Pd catalysts for deep oxidation of benzene. *Microporous Mesoporous Mater.* **2008**, *113*, 472–480. [CrossRef]

41. Yang, P.; Li, J.R.; Zuo, S.F. Promoting effects of Ce and Pt addition on the destructive performances of $V_2O_5/\gamma\text{-}Al_2O_3$ for catalytic combustion of benzene. *Appl. Catal. A Gen.* **2017**, *542*, 38–46. [CrossRef]

42. Li, J.R.; Zuo, S.F.; Qi, C.Z. Preparation and high performance of rare earth modified Co/USY for benzene catalytic combustion. *Catal. Commun.* **2017**, *91*, 30–33. [CrossRef]

43. Zou, X.L.; Rou, Z.B.; Song, S.Q.; Ji, H.B. Enhanced menthane combustion performance over $NiAl_2O_4$-interface promoted $Pd/\gamma\text{-}Al_2O_3$. *J. Catal.* **2016**, *338*, 192–201. [CrossRef]

44. Tang, W.X.; Deng, Y.Z.; Chen, Y.F. Promoting effect of acid treatment on Pd-Ni/SBA-15 catalyst for complete oxidation of gaseous benzene. *Catal. Commun.* **2017**, *89*, 86–90. [CrossRef]

45. Shim, W.G.; Kim, S.C. Heterogeneous adsorption and catalytic oxidation of benzene, toluene and xylene over spent and chemically regenerated platinum catalyst supported on activated carbon. *Appl. Surf. Sci.* **2010**, *256*, 5566–5571. [CrossRef]

46. Bottazzi, G.; Martinez, M.; Costa, M.; Anunziata, O.; Beltramone, A. Inhibition of the hydrogenation of tetralin by nitrogen and sulfur compounds over Ir/SBA-16. *Appl. Catal. A Gen.* **2011**, *404*, 30–38. [CrossRef]

47. Zhao, Q.; Xu, Y.; Li, Y.; Jiang, T.; Li, C.; Yin, H. Effect of the Si/Ce molar ratio on the textural properties of rare earth element cerium incorporated mesoporous molecular sieves obtained room temperature. *Appl. Surf. Sci.* **2009**, *255*, 9425–9429. [CrossRef]

48. Somanathan, T.; Pandurangan, A.; Sathiyamoorthy, D. Catalytic influence of mesoporous Co-MCM-41 molecular sieves for the synthesis of SWNTs via CVD method. *J. Mol. Catal. A Chem.* **2006**, *256*, 193–199. [CrossRef]

49. Bing, J.S.; Li, L.S.; Lan, B.Y.; Liao, G.Z.; Zeng, J.Y.; Zhang, Q.Y.; Li, X.K. Synthesis of cerium-doped MCM-41 for ozonation of p-chlorobenzoic acid in aqueous solution. *Appl. Catal. B Environ.* **2012**, *115–116*, 16–24. [CrossRef]

 *materials*

Article

Microwave-Assisted Pillaring of a Montmorillonite with Al-Polycations in Concentrated Media

Beatriz González [1], Alba Helena Pérez [1], Raquel Trujillano [1], Antonio Gil [2] and Miguel A. Vicente [1,*]

[1] Departamento de Química Inorgánica, Universidad de Salamanca, 37008 Salamanca, Spain; bei@usal.es (B.G.); ahpj@usal.es (A.H.P.); rakel@usal.es (R.T.)
[2] Departamento de Química Aplicada, Universidad Pública de Navarra, 31006 Pamplona, Spain; andoni@unavarra.es
* Correspondence: mavicente@usal.es; Tel.: +34-6-7055-8392

Received: 5 July 2017; Accepted: 29 July 2017; Published: 1 August 2017

Abstract: A montmorillonite has been intercalated with Al^{3+} polycations, using concentrated solutions and clay mineral dispersions. The reaction has been assisted by microwave radiation, yielding new intercalated solids and leading to Al-pillared solids after their calcination at 500 °C. The solids were characterized by elemental chemical analysis, X-ray diffraction, FTIR spectroscopy, thermal analyses, and nitrogen adsorption. The evolution of the properties of the materials was discussed as a function of the preparation conditions. Microwave treatment for 2.5 min provided correctly pillared solids.

Keywords: Al-PILC; Keggin polycation; concentrated media; microwave radiation; pillared montmorillonite

---

1. Introduction

Clay minerals' importance in the field of catalysis is booming, because they have high specific surface area and active acid centers that lead to their use in various reactions such as gasoline desulfurization, terpene isomerization, olefin polymerization, cracking, and numerous fine chemical reactions [1].

One of the most studied groups of clay minerals is that of smectites, which are 2:1 phyllosilicates. The expansibility of the smectite interlayer space depends on four fundamental factors: the nature of the exchangeable, charge-compensating cations, the density of the surface charge and the location of the charge. The best known smectite is montmorillonite, very abundant in nature and with a great variety of properties and applications. Montmorillonites are formed by silicates of Al, Mg, or Fe with various degrees of hydration and amounts of alkaline and alkaline earth exchangeable cations, with a cation exchange capacity (CEC) between 0.3 and 0.8 meq/g [2]. Their basal space ranges from 9.7 Å when the sheets are as close as possible, to 12–14 Å in natural samples, when the interlayer cations are hydrated by a monolayer or a bilayer of water molecules, and to about 23 Å when intercalating voluminous cations. Montmorillonite is one of the clay minerals with higher industrial interest, being easy to obtain at low cost.

Pillared clays (PILC) are a family of molecular engineered porous solids very studied in the last decades. Its preparation is based on the exchange of interlayer cations of the natural clay mineral by voluminous inorganic polycations, obtained by polymerization of various multivalent cations (Al^{3+}, Ga^{3+}, Ti^{4+}, Zr^{4+}, Fe^{3+}, among others). The resulting solids are calcined at moderate temperatures, up to 500 °C, leading to stable materials with a larger spacing than in the initial solid. During calcination, polycations are converted into metal oxides, pillars, inserted between the sheets of the clay mineral and keeping them separated from each other, which prevents the collapse of the structure [3–6]. The most commonly used polycation in the pillaring process is the Keggin polycation, $[Al_{13}O_4(OH)_{24}(H_2O)_{12}]^{7+}$, due to its stability and simple preparation [7–9].

In addition to the nature of the polycation, there are other factors that condition the intercalation process and that must be considered when planning a preparation, such as intercalation time, hydrolysis conditions, and aging of the polycations. Usually, dilute solutions of the polycations and dilute dispersions of the clay minerals are used, looking for optimal conditions for the polymerization of the cations and for the cation exchange reaction required for substituting the compensating cations of the clay mineral by the polycations. Logically, this involves the use of large volumes of water, strongly hindering the scaling-up of this process to prepare large amounts of pillared solids. Some papers have proposed procedures to solve this drawback; these papers are reviewed in [6].

Microwave-hydrothermal treatment has been widely used in recent years in materials science, making the preparation reactions faster than using traditional methods. In clay science, microwave-assisted methods have been used for the preparation of clay minerals, mainly saponite [10], and of Layered Double Hydroxides (LDH) [11], and also for the preparation of PILC [6,12].

The aim of this work is to obtain montmorillonites pillared with aluminum Keggin polycations using microwave radiation and concentrated media—that is, low water volumes—as well as to determine the structural and textural properties of the solids thus obtained.

2. Materials and Methods

2.1. Preparation of the Solids

The clay mineral used was a raw montmorillonite from Cheto, Arizona, USA (from The Clay Minerals Repository, where this sample is denoted as SAz-1). The natural clay mineral was purified before its use by dispersion–decantation, separating the fraction lower than 2 μm. In the present work, this clay mineral is designated as 'RMt'. Its cation exchange capacity was 0.67 meq/g, its basal spacing was 13.60 Å, and its BET specific surface area was 49 m^2/g [13]. For comparison, this solid was calcined at 500 °C, the same temperature as the pillared solids, denoted as 'Mt'.

Four samples were synthesized, varying the aluminum concentration and the treatment time under microwave irradiation. In a typical experiment, the Al-polycation solution was prepared by dissolution of 10 mmol of $AlCl_3 \cdot 6H_2O$ in 20 cm^3 of water and the subsequent slow addition, under vigorous stirring of 22 mmol of NaOH until pH = 4.2 was reached, the condition for the formation of Al_{13}^{7+} polycations. This intercalating solution was added dropwise to a previously prepared montmorillonite dispersion (2 g in 20 cm^3), with an Al/montmorillonite ratio of 5 mmol/g, and it was submitted to microwave treatment at 150 °C several times. For that, the suspensions were sealed in 100 cm^3 Teflon reactors and treated in a Milestone Ethos Plus microwave furnace. The heating process was programmed with EasyWAVE Software. The furnace has a power of 600 W, and initially applies the power needed for a heating ramp of 5 °C/min to the treatment temperature and then to maintain the temperature for the time required. Then, the solids were separated and washed by centrifugation, dried overnight at 70 °C, and finally calcined at 500 °C for 2 h, with a heating rate of 1 °C/min. The conditions of preparation and the names of the solids are summarized in Table 1. All reagents used were supplied by Panreac (Barcelona, Spain), being compounds of the highest purity, and were used without purification.

Table 1. Nomenclature of the solids and preparation conditions.

Name	Dispersion Concentration (g/cm^3)	Intercalating Solution Concentration (mmol Al^{3+}/cm^3)	Microwave Treatment Time (min)
MtAl2.5	2/20	10/20	2.5
MtAl5A	2/20	10/20	5
MtAl5B	2/10	10/10	5
MtAl10	2/20	10/20	10

2.2. Characterization Techniques

Element chemical analyses were carried out at *Servicio General de Análisis Químico Aplicado* (University of Salamanca, Salamanca, Spain), using Inductively Coupled Plasma-Atomic Emission Spectrometry (ICP-AES). X-ray diffraction (XRD) patterns were recorded between 2° and 65° (2θ) over non-oriented powder samples, at a scanning speed of 2° (2θ)/min, by using a Siemens D-500 diffractometer (Siemens España, Madrid, Spain), operating at 40 kV and 30 mA, using filtered Cu Kα radiation (λ = 1.5418 Å). FT-IR spectra were recorded between 450 and 4000 cm^{-1} using a PerkinElmer Spectrum-One spectrometer (Waltham, MA, USA) by the KBr pellet method, with a sample:KBr ratio of 1:300. Thermal analyses were performed on a SDT Q600 TA instrument (New Castle, PA, USA), TG and DTA were carried out simultaneously; all measurements were carried out under a flow of 20 cm^3/min of oxygen (Air Liquide, Madrid, Spain, 99.999%) and a temperature heating rate of 10 °C/min from room temperature to 900 °C. Textural properties were determined from nitrogen (Air Liquide, 99.999%) adsorption–desorption data, obtained at −196 °C using a Micrometrics Gemini VII 2390t (Norcross, GA, USA), Surface Area and Porosity apparatus. Specific surface area (SSA) was obtained by the BET method, external surface area and micropore volume by means of the *t*-method, and the total pore volume from the nitrogen adsorbed at a relative pressure of 0.95 [14].

3. Results and Discussion

Characterization or raw montmorillonite has been reported elsewhere [13], it is a very pure clay mineral.

The elemental chemical compositions of all the Al-PILC samples are given in Table 2. For better assessment of the chemical effects of the treatments, a double normalization was carried out. First, the composition was given as for water-free solids, that is, the metal oxide content was normalized to sum 100%, the effect of the content of water in each solid was thus avoided. Then, the compositions were referred to the content of SiO$_2$ in the original Mt, as the tetrahedral sheet of the montmorillonite was not expected to be affected by the intercalation and pillaring treatment, the amount of SiO$_2$ thus remaining constant, and it can be used as an 'internal standard'. The compositions thus normalized are given in Table 3.

Table 2. Chemical composition of the solids, expressed in content of their metallic oxides (mass percentage).

Sample	SiO$_2$	Al$_2$O$_3$	Fe$_2$O$_3$	MnO	MgO	CaO	Na$_2$O	K$_2$O	TiO$_2$	Total
RMt	55.80	15.92	1.41	0.04	5.58	1.69	0.06	0.06	0.21	80.77
MtAl2.5	49.11	24.37	1.21	0.02	4.10	0.05	0.35	0.13	0.18	79.52
MtAl5A	53.76	28.25	1.35	0.02	5.21	0.04	0.37	0.12	0.20	89.32
MtAl5B	55.45	28.04	1.37	0.03	4.96	0.03	0.36	0.04	0.20	90.48
MtAl10	53.89	28.13	1.28	0.03	4.96	0.04	0.35	0.10	0.19	88.97

Table 3. Chemical composition of the water-free solids, normalized to the SiO$_2$ content in the raw montmorillonite (mass percentage).

Sample	SiO$_2$	Al$_2$O$_3$	Fe$_2$O$_3$	MnO	MgO	CaO	Na$_2$O	K$_2$O	TiO$_2$
Mt	69.09	19.71	1.75	0.05	6.91	2.09	0.07	0.07	0.26
MtAl2.5	69.09	34.29	1.70	0.03	5.77	0.07	0.49	0.18	0.26
MtAl5A	69.09	36.31	1.73	0.02	6.69	0.05	0.47	0.15	0.25
MtAl5B	69.09	34.94	1.70	0.03	6.18	0.03	0.45	0.05	0.25
MtAl10	69.09	36.07	1.64	0.03	6.35	0.05	0.44	0.13	0.24

The amount of aluminum fixed, once the amount existing in the raw montmorillonite was subtracted, was very similar in all the solids, between 14.58% and 16.60% (expressed as Al$_2$O$_3$), enough to intercalate the montmorillonite compensating its CEC. In fact, the polycations were incorporated into the clay mineral structure by cationic exchange of Ca^{2+}, its main exchangeable cation, which was almost completely removed, while K^+ content remained almost constant, suggesting that it was in the raw montmorillonite as feldspar, but not as exchangeable cations. The amount of Fe_2O_3 remained essentially constant, while MgO slightly decreased, suggesting that octahedral Mg^{2+} was scarcely dissolved. The polymerization of Al^{3+} required relatively acid conditions (pH = 4.2), but the contact of the clay mineral with the polycation solution did not significantly alter the composition of the Mt layers.

The powder X-ray diffractograms of pillared solids are shown in Figure 1 and the basal spacing of all the solids are summarized in Table 4. The raw solid was a well-ordered Mt with a basal spacing of 13.90 Å, which collapsed to 9.71 Å after calcination at 500 °C [13]. The treatment with the intercalating solutions always led to the expansion of the interlayer region, the basal spacing varying between 15.87 Å and 18.14 Å for the intercalated, dried solids (Table 4). The maximum value, 18.14 Å, similar to the values usually reported under conventional pillaring, was observed for MtAl5A solid, implying a height of the interlayer region of approximately 8.4 Å, which coincided with the height of the Al_{13}^{7+} polycations. The other three solids did not reach this value, remaining around 16 Å, which supposes a height of the interlayer region of ~6.3 Å, suggesting that the polymerization was not optimal or that the previously formed polycations suffered some changes in the degree of polymerization during their addition to the clay mineral or during the microwave treatment.

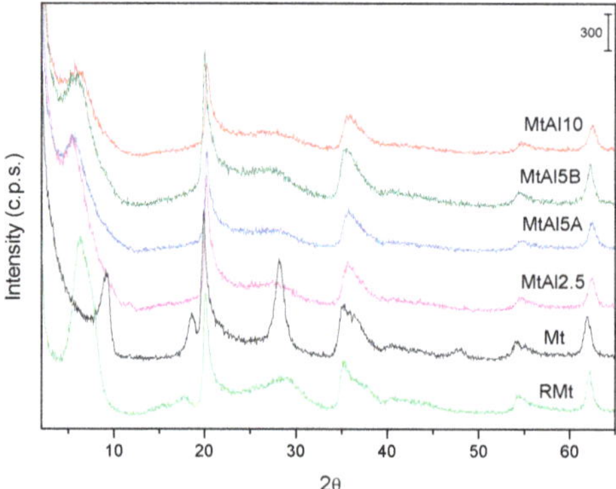

Figure 1. Powder X-ray diffractograms of the pillared solids calcined at 500 °C. For comparison, the diffractograms of the raw montmorillonite (RMt) and of this sample calcined at 500 °C (Mt) are also given.

Table 4. Basal spacing (Å) for intercalated and calcined samples.

Sample	Intercalated	500 °C
Mt	13.90 *	9.71
MtAl2.5	16.14	16.50
MtAl5A	18.14	15.58
MtAl5B	15.87	15.56
MtAl10	16.22	15.58

* For Mt sample, basal spacing of the raw montmorillonite.

In general, the basal spacing decreased during the calcination process. Calcination of intercalated samples usually produces a decrease in their basal spacing by dehydration and dehydroxylation of the Al_{13}^{7+} polycations until the formation of clusters close to Al_2O_3, and this is the trend found for MtAl5A, MtAl5B, and MtAl10 solids. However, in the case of the MtAl2.5, there was an increase in the basal spacing during calcination. This sample had been submitted to a lower time of treatment with microwave radiation, so it seems that the intercalation had not been completed during this treatment and the process could be completed at the beginning of the calcination.

The effects due to in-layer reflections, independent of *c*-stacking, were recorded at the same positions for all the solids, indicating that the layers were not modified. No reflections belonging to crystalline Al_2O_3 were observed.

The FWHM (Full Width at Half Maximum) index of the basal 001 reflection varied from the intercalated to the calcined solids (Table 5), showing changes in the long-distance arrangement according to this plane, i.e., the number of layers correctly stacked along the *c* axis. Raw montmorillonite had a relatively small FWHM index, probably due to the ordering caused by the dispersion–decantation purification procedure (when calcining at 500 °C, a very sharp peak, with FWHM of only 0.999° was obtained, due to the collapse to TOT arrangement). The intercalation tended to order the sheets, that is, the interaction with the polycations made more layers to correctly stack, due to the successive interaction layers–polycations. Calcination for forming the final pillared solids increased or decreased the value of FWHM index, depending of the considered solid. Usually, calcination causes a decrease of ordering, and this actually occurred in MtAl5B and MtAl10, however, the opposite behavior was observed for MtAl2.5 and MtAl5A. This may be due to the high speed of the intercalation procedure in these solids, which may not allow the correct stacking of the layers, by which the stacking may continue during the first steps of the calcination.

Table 5. Value of FWHM index of the solids.

Solid	FWHM/°
RMt	2.372
Mt-500	0.999
MtAl2.5	1.929
MtAl2.5-500	1.775
MtAl5A	1.768
MtAl5A-500	1.701
MtAl5B	1.701
MtAl5B-500	2.290
MtAl10	1.607
MtAl10-500	1.887

The FT-IR spectra of the all samples were similar to that of natural Mt (Figure 2). The O–H stretching band was recorded close to 3440 cm^{-1} and the H–O–H bending vibrational mode at 1637 cm^{-1}. Bands corresponding to the Si–O–Si and Si–O–Al vibrations were recorded at 1036 and 522 cm^{-1}. No Al–O bands were observed for the intercalated clays. The intensity of the O–H stretching mode of the hydroxyls bonded to the metals, recorded around 3600 cm^{-1}, increased after intercalation in MtAl10, suggesting an interaction with the polycations, and an increase in the acidity of this solid.

The thermogravimetric curve of the intercalated solids (that from MtAl10 is included in Figure 3 as an example) showed a ~12% mass loss at low temperature, associated with an endothermic effect centered at 94 °C in the DTA curve. This effect was due to the removal of water located in the interlayer region, bonded to the polycations, or adsorbed on the external surface of the sheets. In the raw montmorillonite, this effect produced an 11% mass loss, with an endothermic associated effect centered at 100 °C. Thus, the loss of the exchangeable cations and the cations coordinated to them was compensated by the water adsorbed on the Al_{13}^{7+} polycations, giving up to similar effects for both solids. In the central temperature range, a significant mass loss effect (~6%) was observed in the pillared solids, with three slight inflexions, associated with an exothermic effect centered at 242 °C and a shoulder at 396 °C. These effects are not observed in the raw montmorillonite, and should be due to the dehydration of the polycations, confirming their correct intercalation on montmorillonite interlayer.

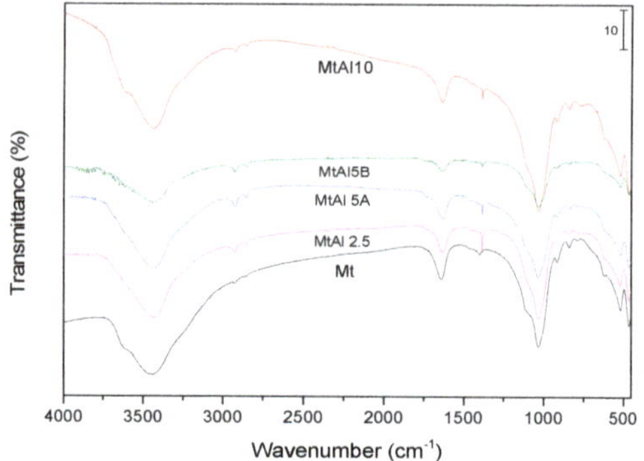

Figure 2. FT-IR spectra of the intercalated solids.

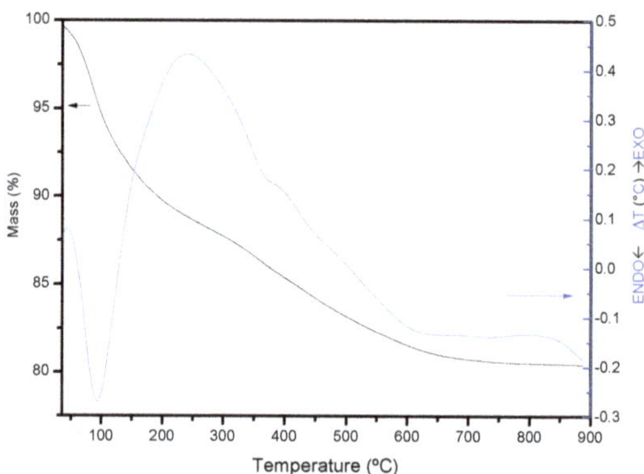

Figure 3. Thermal curves, TG, and DTA of MtAl10 solid.

The textural properties were studied by N_2 adsorption–desorption experiments. The adsorption isotherms (Figure 4) belonged to type II from IUPAC classification, with a H4 type hysteresis loop at high relative pressures, a form associated with narrow slit pores [15]. The loop had an inflexion at a relative pressure value of 0.4, being reversible at low p/p° values for the non-calcined solids, indicating that pores were not rigid but flexible, and became practically reversible for the calcined solids.

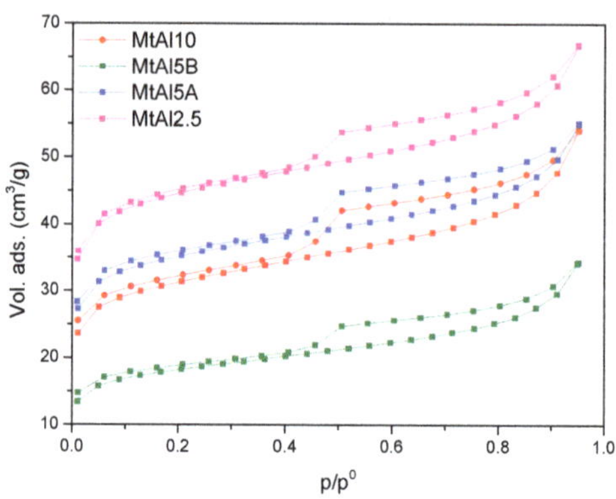

Figure 4. Nitrogen adsorption–desorption isotherms of the pillared solids.

BET specific surface area, external surface area, and micropore volume data are summarized in Table 6. Intercalated solids had specific surface areas higher than the starting clay mineral, with values between 107 and 137 m^2/g. The intercalation of the Al$_{13}^{7+}$ polycations produced a remarkable separation of the sheets, which would tend to increase the specific surface; however the interlayer region was occupied by the polycations, so the nitrogen molecules could not freely access it. Both effects compensated among them, but the first was predominant. The porosity ranged between 0.039 and 0.054 cm^3/g and the external surface was in the range of 36–45 m^2/g (~28–37% of the total surface area).

Table 6. Specific surface area (S$_{BET}$), external surface area (S$_{ext}$) and micropore volume (V$_m$) of natural montmorillonite [13] and of the pillared solids.

Sample	S$_{BET}$ (m^2/g)	S$_t$ (m^2/g) *	V$_m$ (cm^3/g)
RMt	49	49 (100)	0.000
Mt-500	80	64 (80)	0.009
MtAl2.5	123	45 (37)	0.044
MtAl2.5-500	135	49 (36)	0.047
MtAl5A	137	39 (28)	0.054
MtAl5A-500	107	43 (40)	0.035
MtAl5B	107	37 (35)	0.039
MtAl5B-500	57	27 (47)	0.016
MtAl10	115	36 (31)	0.044
MtAl10-500	96	43 (45)	0.030

* In brackets, percentage of external surface to the total specific surface area.

Calcination produced a decrease in the S$_{BET}$ for MtAl5A, MtAl5B, and MtAl10 solids, which aligned with the expected trend, a decrease of this magnitude due to the collapse of the sheets, while the external surface remained practically constant, increasing its relative contribution to the total surface, up to 47%. Again, MtAl2.5 showed opposite behavior, with the S$_{BET}$ increasing upon calcination. As previously commented, this was the solid submitted to a shorter time of microwave treatment. The increase in surface area again suggested that the pillaring procedure was completed during the calcination step.

The overall analysis of these results indicated that Al-PILC can be obtained by microwave treatment of concentrated solutions and dispersions for a time as short as 2.5 min, although in this case, the pillaring process seemed to be completed during the final calcination step of the preparation procedure. Thus, microwave radiation should strongly facilitate the preparation of these solids in large amounts.

4. Conclusions

Montmorillonite was effectively pillared with Al$_{13}^{7+}$ polycations, using microwave radiation and small volumes of concentrated solutions of the polycations and concentrated dispersions of the clay mineral. The solids obtained showed structural characteristics comparable to Al-PILC prepared from the classical method. Thus, the use of microwave radiation allows a significant reduction in the times of treatment and the volumes of solutions/dispersions required by the conventional preparation method.

Acknowledgments: Research projects MAT2013-47811-C2-1-R and MAT2016-78863-C2-R, jointly financed from the Spanish Agencia Estatal de Investigación, AEI (Spanish Ministry of Economy and Competitiveness, MINECO) and the European Regional Development Fund (ERDF).

Author Contributions: All the authors conceived, designed, and performed the experiments, analyzed the data, and wrote the paper.

Conflicts of Interest: The authors declare no conflict of interest.

References

1. Bergaya, F.; Lagaly, G. (Eds.) *Handbook of Clay Science*, 2nd ed.; Elsevier: Amsterdam, The Netherlands, 2013.
2. Nesse, W.D. *Introduction to Mineralogy of Clays: Clays and the Environment*; Springer: Berlin, Germany, 1995.
3. Gil, A.; Korili, S.A.; Trujillano, R.; Vicente, M.A. (Eds.) *Pillared Clays and Related Catalysts*; Springer: Berlin, Germany, 2010.
4. Gil, A.; Gandía, L.M.; Vicente, M.A. Recent Advances in the Synthesis and Catalytic Applications of Pillared Clays. *Catal. Rev.* **2000**, *42*, 145–212. [CrossRef]
5. Gil, A.; Korili, S.A.; Vicente, M.A. Recent Advances in the Control and Characterization of the Porous Structure of Pillared Clay Catalysts. *Catal. Rev.* **2008**, *50*, 153–221. [CrossRef]
6. Vicente, M.A.; Gil, A.; Bergaya, F. Pillared Clays and Clay Minerals. In *Handbook of Clay Science*, 2nd ed.; Bergaya, F., Lagaly, G., Eds.; Elsevier: Amsterdam, The Netherlands, 2013; Chapter 10.5.
7. Johansson, G. On the Crystal Structure of Some Basic Aluminium Salts. *Acta Chem. Scand.* **1960**, *14*, 771–773. [CrossRef]
8. Lahav, N.; Shani, U.; Shabtai, J. Cross-Linked Smectites. I. Synthesis and Properties of Hydroxy-Aluminum-Montmorillonite. *Clays Clay Miner.* **1978**, *26*, 107–115. [CrossRef]
9. Bottero, J.Y.; Cases, J.M.; Fiessinger, F.; Poirier, J.E. Studies of hydrolyzed aluminum chloride solutions. 1. Nature of aluminum species and composition of aqueous solutions. *J. Phys. Chem.* **1980**, *84*, 2933–2939. [CrossRef]
10. Jaber, M.; Komarneni, S.; Zhou, C.-H. Synthesis of Clay Minerals. In *Handbook of Clay Science*, 2nd ed.; Bergaya, F., Lagaly, G., Eds.; Elsevier: Amsterdam, The Netherlands, 2013; Chapter 7.2.
11. Forano, C.; Costantino, U.; Prévot, V.; Taviot Gueho, C. Layered Double Hydroxides (LDH). In *Handbook of Clay Science*, 2nd ed.; Bergaya, F., Lagaly, G., Eds.; Elsevier: Amsterdam, The Netherlands, 2013; Chapter 14.1.
12. Fetter, G.; Bosch, P. Microwave effect on Clay Pillaring. In *Pillared Clays and Related Catalysts*; Gil, A., Korili, S.A., Trujillano, R., Vicente, M.A., Eds.; Springer: Berlin, Germany, 2010; Chapter 1.
13. González-Rodríguez, B.; Trujillano, R.; Rives, V.; Vicente, M.A.; Gil, A.; Korili, S.A. Structural, textural and acidic properties of Cu-, Fe- and Cr-doped Ti-pillared montmorillonites. *Appl. Clay Sci.* **2015**, *118*, 124–130. [CrossRef]
14. Rouquerol, F.; Rouquerol, J.; Sing, K. *Adsorption by Powders and Porous Solids—Principles, Methodology and Applications*; Academic Press: Cambridge, MA, USA, 1998.
15. Sing, K.S.W.; Everett, D.H.; Haul, R.A.W.; Moscou, L.; Pierotti, R.A.; Rouquerol, J.; Siemieniewska, T. Reporting physisorption data for gas/solid systems with special reference to the determination of surface area and porosity. *Pure Appl. Chem.* **1985**, *57*, 603–619. [CrossRef]

Article

Scale up Pillaring: A Study of the Parameters that Influence the Process

Francine Bertella and Sibele B. C. Pergher *

Laboratory of Molecular Sieves—LABPEMOL, Institute of Chemistry, Federal University of Rio Grande do Norte—UFRN, Av. Senador Salgado Filho, 3000, Lagoa Nova, University Campus, 59072-970 Natal-RN, Brazil; francinebertella@gmail.com
* Correspondence: sibelepergher@gmail.com; Tel.: +55-84-9413-5418

Academic Editor: Miguel A. Vicente
Received: 28 April 2017; Accepted: 14 June 2017; Published: 27 June 2017

Abstract: Pillared clays (PILCs) are interesting materials mostly due to their high basal spacing and surface area, which make them suitable for adsorption and catalysis applications, for example. However, the production of these materials on industrial scale is dependent on research about what parameters influence the process. Thus, the objective of this work was to evaluate what parameters influence the pillaring procedure. For this, pillared clays were synthesized following three series of experiments. In the first series, the effect of the amount of water in a clay suspension was evaluated. The best results were obtained by using diluted suspensions (1 g of clay to 100 mL of water). In the second series, several pillaring methods were tested. In the third series, the amount of pillared clay was raised to 50 g. Fifty grams of pillared clay can be obtained using the pillaring agent synthesized at 60 °C with further aging for 24 h, and this material exhibited high basal spacing (17.6 Å) and surface area (233 m^2/g). These values are comparable with the traditional pillaring method using only 3 g of clay.

Keywords: clays; Al-PILC; pillared clays; scale up

---

1. Introduction

Clays are abundant natural materials with a low cost. To improve the catalytic and adsorptive properties of these materials, clays can be submitted to different processes, such as pillaring and acid activation. Usually, clay pillaring occurs through cation exchange of natural cations (Ca^{2+}, Na^+, Mg^{2+}) present between clay layers for bigger cations such as polyhydroxy cations of Al (a pillaring agent). These larger cations act as pillars, separating the clay layers, increasing the basal space and creating microporosity. The calcination provides stability to the pillared clay, and the resultant material presents small cavities and a large surface area. These properties, combined with the low cost of clay, make pillared clays alternative catalysts to zeolites.

The majority of articles found in the literature use polyhydroxy cations (mostly aluminum, zirconium, iron, and titanium) as pillaring agents [1–3]. These cations can be used separately (just one type of pillar) [4–10] or as mixed pillars [11–17].

However, most papers about pillared clays report the use of polyhydroxy cations of only aluminum as the pillaring agent. Solutions containing this complex are made through addition of a base (hydroxide or carbonate) to a salt of aluminum ($AlCl_3$ or $Al(NO_3)_3$) until a molar ratio of OH/Al = 2.5 is attained or through dissolution of powdered Al in $AlCl_3$ [1]. Solutions prepared with these methods contain mostly three species: hydrated monomeric aluminum, the polyhydroxy cation $[Al_{13}O_4(OH)_{24}(H_2O)_{12}]^{7+}$, also known as ε-Keggin ion or Al_{13}, and polynuclear aluminum [18–20]. Some of these Al-pillared clays have been reported as support for active phases, such as metals, in catalytic reactions [21–23].

Variations in the synthesis provide different characteristics to the resulting materials, and the most important parameters that can be varied are related to the formation, intercalation, and posterior fixation of the polynuclear cations between the clay layers. No general rules exist about the best conditions for the synthesis [24].

Several pillaring methods and pillaring solutions have been reported in the literature. Despite the major studies that have been performed on a laboratory scale (with small quantities of clay), some authors have utilized concentrated suspensions to increase the scale of pillaring [14,25–31]. In this sense, to reduce the volume of water and intercalation times, Sanabria et al. [14] have performed pillaring procedures by using dialysis membranes and ultrasound to produce pillared clays with mixed Al-Fe, Al-Ce-Fe pillars. By this methodology, they were able to produce pillared clays with similar features to those of the solid synthesized by their conventional procedure.

Nevertheless, a closer look into the literature reveals that an enormous amount of parameters can influence the pillaring processes, like the starting clay material, the synthesis condition and nature of the pillaring agent, the use of concentrated or dilute clay suspensions, time and temperature of intercalation and conditions of washing, drying and calcining processes. In this view, to study a scale-up pillaring process it is of primary importance the optimization of key parameters in order to obtain pillared clays with comparable features (basal spacing and surface areas) similar to the material synthesized by the conventional procedure. To achieve this purpose, a systematic study of the most important variables that influence a scale-up pillaring process, like pillaring method, concentration of clay suspension and amount of clay, using the same starting clay is of paramount importance.

Thus, the aim of the present study is to prepare and characterize pillared clays modifying the conditions of the synthesis to determine which parameters influence the increase in the pillaring scale of a bentonite clay.

2. Materials and Methods

2.1. Traditional Pillaring Method

A bentonite clay consisting of mostly montmorillonite, supplied by Colorminas Colorifício e Mineração S/A, Brazil, was used in this study. The chemical composition of the pristine clay is: 83.02% Si, 13.55% Al, 1.01% Ca, 1.67% Mg, 0.25% Na and 0.50% K. The cation exchange capacity (CEC) determined for the natural clay is 155 mEq/100 g.

The first pillaring stage is the preparation of the pillaring solution. In this way, 500 mL of a NaOH solution (Sigma-Aldrich, St. Louis, MO, USA) and 250 mL of a $AlCl_3 \cdot 6H_2O$ solution (Vetec Fine Chemicals, Duque de Caxias, Brazil), both 0.2 mol/L, were used in order to prepare the pillaring agent. The NaOH solution was dripped slowly into the Al solution (approx. 1 mL/min) under constant stirring at ambient temperature. Then, the pillaring solution was maintained under these synthesis conditions for 6 days. The OH/Al ratio was 2, and 15 mEq Al/g of clay was used [2].

After the 6 days of aging the pillaring solution, 3 g of clay was stirred into 300 mL of distilled water (1 g/100 mL) for 2 h at room temperature to hydrate the interlayer cations and expand the lamellae.

Subsequently, the pillaring agent was added to the clay suspension, and the mixture was stirred for 2 h at room temperature to allow the natural clay cations to be exchanged with the prepared polyhydroxy cations (pillaring agent). The material was vacuum-filtered (using a Büchner funnel and filter paper serving as the porous barrier), washed abundantly with distilled water, dried overnight in an oven at 60 °C and finally calcined at 450 °C for 3 h in a muffle furnace (the heating rate was 5 °C/min).

This pillaring procedure has been performed several times to prove its reproducibility.

2.2. Scale Up Pillarizations

The experiments were divided into three series. In series 1, the effect of water during the expansion of clay was studied. In series 2, the method of pillaring was studied, and in series 3, the amount of pillared clay was increased.

2.2.1. Series 1: Effect of Water

To evaluate the effect of water utilized to expand the clay, several methods were tested (Table 1). The OH/Al = 2 ratio and 15 mEq Al/g of clay were maintained in all experiments.

Table 1. Modified parameters to evaluate the effect of water.

Method	Solution Concentrations	Relation of Clay/Water	Amount of Pillared Clay
Traditional	0.2 mol/L	1 g/100 mL	3 g
1	0.6 mol/L	1 g/10 mL	10 g
2	0.6 mol/L	1 g/50 mL	10 g
3	0.6 mol/L	1 g/100 mL	10 g
4	0.6 mol/L	1 g/100 mL	10 g
5	0.6 mol/L	1 g/100 mL	10 g

Methods 1, 2, and 3 were performed according to the traditional procedure, only modifying the amount of water present in the clay suspension. However, methods 4 and 5 were performed in a distinct form, with the exception of the synthesis of the pillaring solution. In method 4, after preparation of the clay suspension (10 g of clay with 1 L of distilled water), 500 mL of water was removed by centrifugation (5000 rpm for 10 min). Later, the remaining suspension was added to the pillaring solution, and the mixture was stirred for 2 h. The clay was vacuum-filtered, washed with distilled water, dried in an oven at 60 °C, and calcined at 450 °C for 3 h, as previously described.

Method 5 was performed similarly to method 4, but after the stage of hydration of the interlayer cations, 900 mL of water was removed from the clay suspension by centrifugation. The remaining 100 mL of clay suspension was added to the pillaring agent performing the pillaring stage similarly to the other experiments. This procedure (removal of water suspension) was performed to verify the influence of the amount of water utilized in the clay suspension.

All methods from this series have been performed more than once in order to check their reproducibility.

2.2.2. Series 2: Method Effect

In this series of experiments, several pillaring methods were tested (Table 2). In method 6, 150 mL of a NaOH solution and 75 mL of a $AlCl_3 \cdot 6H_2O$ solution, both 0.6 mol/L, were utilized to prepare the pillaring solution. The NaOH solution was dripped slowly (1 mL/min) into the Al solution under constant stirring at ambient temperature and maintained under these conditions for 6 days. Subsequently, the pillaring agent was transferred to a round bottom flask containing 3 g of dry clay and kept under reflux for 24 h at 80 °C. Then, the material was vacuum-filtered, washed with distilled water, dried in an oven at 60 °C, and calcined at 450 °C as previously described in Section 2.1.

Table 2. Modified parameters to study the effect of the method.

Method	Solution Concentrations	Relation of Clay/Water	Amount of Pillared Clay
Traditional	0.2 mol/L	1 g/100 mL	3 g
6	0.6 mol/L	-	3 g
7	0.6 mol/L	1 g/100 mL	10 g
8	0.6 mol/L	-	10 g
9	0.2 mol/L	1 g/100 mL	3 g
10	0.2 mol/L	1 g/50 mL	3 g
11	0.6 mol/L	1 g/100 mL	10 g

Experiment 7 proceeded as described in the traditional method. However, after 6 days of aging, the pillaring solution was stored for a month in an amber vial to check whether the properties of the pillaring solution remained over the course of time. The pillarization procedure was performed as described in the traditional method.

An in situ pillarization, tested in method 8, was performed differently from all other methods. Ten grams of clay was added to 250 mL of an $AlCl_3 \cdot 6H_2O$ solution (0.6 mol/L) and stirred for 2 h. Subsequently, 500 mL of a NaOH solution (0.6 mol/L) was dripped slowly (1 mL/min) into the Al solution and stirred for 6 days at ambient temperature. After the aging time, the material was washed, filtered, dried, and calcined as previously described in Section 2.1.

Experiment 9 was performed as described previously. However, the clay was first expanded with water (3 g in 300 mL) for 2 h at ambient temperature. After the solution of aluminum chloride (250 mL of 0.44 mol/L) was added to the clay suspension, the mixture was stirred for 2 h. This solution had to be more concentrated to maintain the final concentration at 0.2 mol/L after the solution had been added to the clay suspension. The following procedures were performed as described in the previous method.

Pillaring method 10 was performed as described by Leite et al. [32]. To prepare the pillaring agent, 500 mL of a NaOH solution and 250 mL of an $AlCl_3 \cdot 6H_2O$ solution, both 0.2 mol/L, were utilized. However, during the dripping (1 mL/min), the solution was kept at 60 °C with stirring. Then, the solution was stirred at ambient temperature for 24 h. Subsequently, 3 g of clay was dispersed in 150 mL of water for 48 h of stirring at ambient temperature. Then, the pillaring agent was added to the clay suspension and stirred for 48 h more. Finally, the clay was filtered, washed, dried, and calcined as previously described in Section 2.1.

Method 11 was performed in the same manner as method 10. However, the stirring time was reduced from 48 to 2 h. The same procedure was performed with the cation exchange stage: the pillaring solution and the clay suspension were kept in contact for just 2 h. The pillaring agent synthesis was performed as described in method 10, without modifications.

The majority of methods from this series have been performed just once.

2.2.3. Series 3: Increase in the Amount of Pillared Clay

In this series of experiments, the amount of pillared clay was increased. Table 3 presents the modified parameters for each experiment. In all methods, 1 g of clay to 100 mL of water was utilized.

Table 3. Modified parameters to increase the amount of pillared clay.

Method	Solution Concentrations	Amount of Pillared Clay	Observation
Traditional	0.2 mol/L	3 g	Pil. Ag. 6 days [a]
3	0.6 mol/L	10 g	Pil. Ag. 6 days
11	0.6 mol/L	10 g	Pil. Ag. 60 °C, 24 h [b]
12	1.2 mol/L	20 g	Pil. Ag. 6 days
13	1.2 mol/L	20 g	Pil. Ag. 60 °C, 24 h
14	1.5 mol/L	50 g	Pil. Ag. 6 days
15	1.2 mol/L	50 g	Pil. Ag. 6 days
16	1.2 mol/L	50 g	Pil. Ag. 60 °C, 24 h

[a] Pillaring agent prepared over 6 days; [b] pillaring agent prepared at 60 °C in 24 h.

Pillaring methods 3, 12, 14 and 15 were performed in the traditional manner: the pillaring agent was stirred for 6 days at ambient temperature. The difference between the methods is the concentration of the solution to maintain the relation of 15 mEq Al/g of clay.

In method 14, solutions of 1.5 mol/L were utilized. Due to this high concentration, the pillaring solution became turbid, requiring stirring for 13 days, but even after this period, the pillaring solution still presented some turbidity.

In methods 11, 13 and 16, the pillaring solution was kept at 60 °C during the dripping (1 mL/min) of the sodium hydroxide solution into the aluminum chloride solution. Then, the pillaring agent was stirred at ambient temperature for 24 h. The following procedures were performed as described above.

All methods from this series have been performed more than once in order to check their reproducibility.

2.2.4. Characterization

The prepared materials were characterized by X-ray diffraction (XRD) on a Shimadzu-XRD-7000 apparatus (Shimadzu, Kyoto, Japan) using Cu radiation (λ = 1.54 Å) and 0.02° step size. N_2 physisorption isotherms were measured using a Quantachrome-NOVA 2200e apparatus (Quantachrome, Boynton Beach, FL, USA). Prior to analysis, the samples were degassed for 3 h at 300 °C under vacuum. The surface areas were obtained using the Brunauer, Emmet and Teller (BET) method, and the micropore volumes and external areas were calculated by the t-plot method using the Harkins-Jura-de Boer t-equation [33]. The total pore volume was calculated at a partial pressure p/p_0 of 0.95. The micropore area was calculated as the difference between the BET and external surface areas.

3. Results and Discussion

3.1. Series 1: Effect of Water

Figure 1 shows the X-ray diffraction patterns of the natural clay, the pillared clay using the traditional method and the samples of series 1 (1 to 5). Regarding to the natural clay, the material composition is based mostly on montmorillonite (2θ = 5.8°; 17.7°; 19.8° and 35.1°) and quartz (2θ = 26.6°) [34].

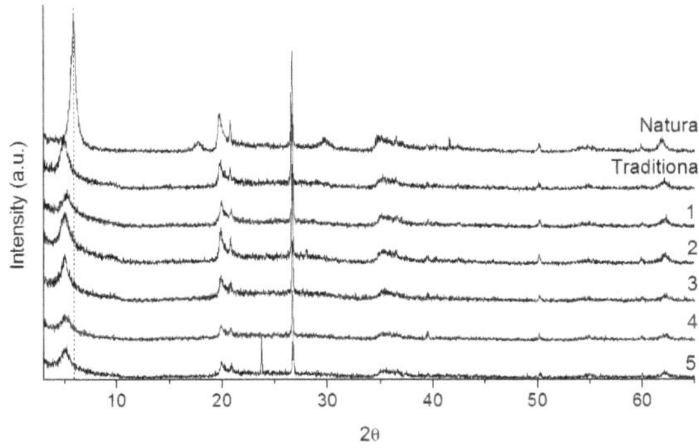

Figure 1. X-ray diffractograms of the natural, traditional and 1 to 5 (series 1) pillared clays.

The shift of the first reflection, which was due to the (001) plane, to smaller angles (2 theta axis) shows that the basal spacing increased (Table 4), indicating that the clay cations were exchanged for the prepared polyhydroxy cations.

The pillared clays using the traditional methods (2 and 3) presented more intense (001) reflections compared to samples 1, 4, and 5; indicating a more organized structure of the lamellae. This more organized structure is reflected in the values of the basal spacing and the surface areas obtained (Table 4). Sample 1 presented the lowest values of surface area (142 m^2/g) and basal spacing (16.8 Å) due to the amount of water used to expand the clay (100 mL). In methods 4 and 5, 500 and 900 mL of water, respectively, were removed before the stage of pillarization. As a consequence of this removal of water, these samples presented lower surface areas (149 m^2/g) in comparison with PILCs obtained by methods 1, 3 and traditional (higher than 200 m^2/g). Therefore, the amount of water used to expand the lamellae also exerts a great influence in the cationic exchange of natural clay cations by the prepared oligomers, translating into a decrease in surface area when fewer amounts of water (methods 4 and 5) are used. Similar results were obtained with samples 2 and 3, indicating that the minimum relation of clay/water to obtain materials with elevated basal space and surface area (above 200 m^2/g) is 1/50.

Figure 2 presents a comparison between the basal spacing, the concentration of the clay suspension and the surface area BET of this series of samples. The best results were obtained through method 3, where elevated amount of water in the clay suspension was used (1/100) and was maintained until the end of the process. For samples 4 and 5, a drastic reduction of surface area occurred due to the removal of water from the clay suspension before the stage of pillaring, which proves that the high quantity of water (diluted suspensions) generates materials with elevated basal spacing and surface areas.

Table 4. Modified parameters, basal spacings and surface area (BET) values of series 1 samples.

Method	Relation of Clay/Water (g/mL)	Mass of Clay	Amount of Water in Clay Suspension	d_{001} (Å)	S_{BET} (m^2/g)	Observations
Natural	-	-	-	15.1	58	-
Traditional	1/100	3 g	300 mL	17.8	234	-
1	1/10	10 g	100 mL	16.8	142	-
2	1/50	10 g	500 mL	17.5	217	-
3	1/100	10 g	1000 mL	17.9	216	-
4	1/100	10 g	1000 mL	17.8	150	−500 mL
5	1/100	10 g	1000 mL	17.3	149	−900 mL

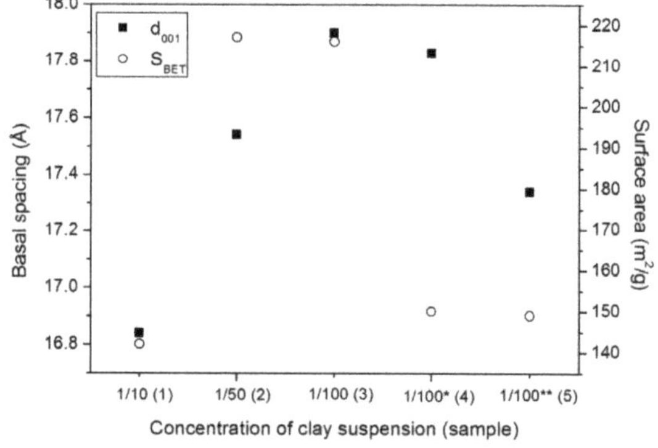

Figure 2. Relation between the concentration of the clay suspension to basal space (Å) and surface area BET (m^2/g) for series 1 samples. * 500 mL and ** 900 mL of water removed before pilarization.

Table 5 presents the textural parameters (BET surface area, micropore area, external area, total pore volume and micropore volume) of series 1 materials.

Table 5. Textural parameters evaluated by N_2 physisorption isotherms of series 1 samples.

Method	S_{BET} (m^2/g)	S_{micro} (m^{2}g)	S_{ext} (m^2/g)	V_{total} (cm^3/g)	V_{micro} (cm^3/g)
Natural	58	19	39	0.070	0.010
Traditional	234	195	39	0.146	0.100
1	142	115	27	0.093	0.059
2	217	180	37	0.135	0.093
3	216	180	36	0.131	0.092
4	150	125	25	0.098	0.064
5	149	120	29	0.101	0.062

When the pillared clay prepared by the traditional method is compared with natural clay, the external area values, which include the external surface plus the meso and macropore area, did not vary. The micropore volume is approximately 10 times larger in the pillared clay due to the increase in basal space. Because of this increase in the micropore volume, the micropore area is approximately

10 times larger. This area is responsible for the increase in total area (BET) in the pillared material. Methods 4 and 5 presented the closest results to those when using the traditional method, with surface areas above 200 m^2/g, a micropore area of 180 m^2/g and a pore volume (total and micro) similar to the related values in the literature [2,35].

3.2. Series 2: Method Effect

In this series, several pillaring methods were evaluated according to basal spacings, and surface areas obtained. The pillaring methods that employed reflux (sample 6) and in situ processes (samples 8 and 9) resulted in low basal spacings and low surface areas, indicating that these methods were not effective (Table 6). Samples 7, 10, and 11 presented larger basal spacings than the natural clay (Figure 3).

Table 6. Modified parameters, basal spacings and surface area values of series 2 methods.

Method	Relation of Clay/Water (g/mL)	Mass of Clay	Amount of Water in Clay Suspension	d_{001} (Å)	S_{BET} (m^2/g)	Observations
Natural	-	-	-	15.1	58	-
Traditional	1/100	3 g	300 mL	17.8	234	-
6	-	3 g	-	13.7	85	Refluxed
7	1/100	10 g	1000 mL	17.2	201	1 month
8	-	10 g	-	17.1	71	In Situ
9	1/100	3 g	300 mL	12.9	74	In Situ w/exp [a]
10	1/50	3 g	150 mL	17.7	225	Pil. Ag. 1 day [b]
11	1/100	10 g	1000 mL	17.5	237	Pil. Ag. 1 day

[a] In situ with expansion; [b] pillaring agent prepared over 1 day.

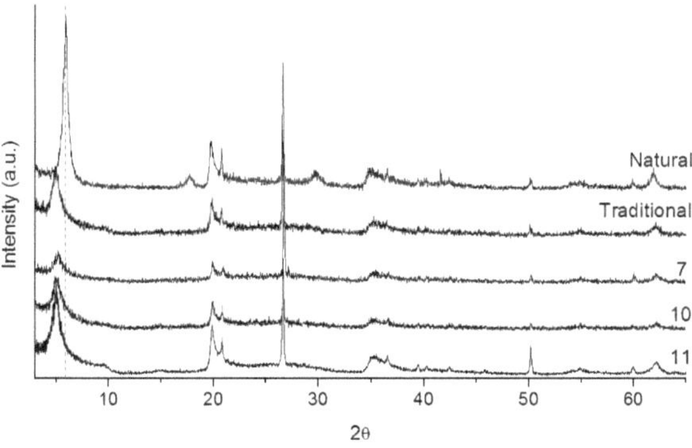

Figure 3. X-ray diffractograms of the natural clay, traditional pillared clay and 7, 10 and 11 pillared clays (series 2).

Method 7, in which the pillaring agent was allowed to stand for a month and used subsequently, resulted in high surface area and basal spacing, indicating that the properties of the pillaring agent had not changed over time, proving that large amounts of pillaring agent could be produced and stored without the pillaring solution losing its function.

Samples 10 and 11, in which the pillaring agent synthesis was performed in one day with heating at 60 °C, presented high basal spacings and surface areas, demonstrating the insertion of Al pillars. Thus, in the methods in which the pillaring agent was prepared separately (without being in situ) and with a clay expansion in water, the best results were obtained. Therefore, the pillaring method influences the characteristics of the materials obtained, and when comparing methods 10 and 11, there is no need to expand the clay lamellae and to submit the clay to cationic exchange for 48 h. These stages are fast, so 2 h is sufficient for each procedure (clay expansion and cationic exchange) to obtain high basal spacings and surface areas.

Table 7 presents the textural parameters obtained using methods 7, 10, and 11, which represent the highest values of basal spacings of the series 2 samples. The samples prepared using methods 7, 10, and 11 are compared to the natural clay and the sample pillared by the traditional method (Table 7).

Table 7. Textural parameters evaluated by N_2 physisorption isotherms of series 2 samples.

Method	S_{BET} (m²/g)	S_{micro} (m²/g)	S_{ext} (m²/g)	V_{total} (cm³/g)	V_{micro} (cm³/g)
Natural	58	19	39	0.070	0.010
Traditional	234	195	39	0.146	0.100
7	201	169	32	0.122	0.087
10	225	184	41	0.140	0.095
11	237	196	41	0.147	0.102

The samples prepared by methods 10 and 11 presented similar features in comparison with the clay pillared by the traditional method. The surface areas obtained by these materials are in the same order (225 and 237 m²/g) than the PILC prepared by the traditional pillaring procedure (234 m²/g). The same trend is observed for the others parameters (micropore and external areas and total and micropore volumes). In fact, the sample pillared by method 11 presented results slightly higher than the PILC obtained by traditional pillaring method, proving that the former method is highly efficient for the synthesis of pillared clays.

Thus, the amount of pillared clay was increased when the methods 3 (6 days pillaring agent) and 11 (1 day pillaring agent) were followed, using the relation of 1 g of clay to 100 mL of water in the clay suspension (higher dilution). This relation was used because, according to the results of series 1, as larger amounts of water are used in the clay suspension, the surface areas and basal spacings become higher.

3.3. Series 3: Increase in the Amount of Pillared Clay

Figure 4 presents the X-ray diffraction patterns of series 3. Pillarization occurred for all methods except method 14, in which a small shift of the (001) reflection to higher angles of the 2θ axis occurred, indicating a smaller basal spacing in relation to the other samples (Table 8).

For all methods, elevated basal spacings (above 17.5 Å) were obtained, except for method 14, where a value of 16.6 Å was calculated. This small value is due to the concentration used to prepare the pillaring agent (1.5 mol/L) because even after stirring for 13 days, the pillaring solution was still turbid, indicating the formation of agglomerates and other species beyond the Keggin ion [19].

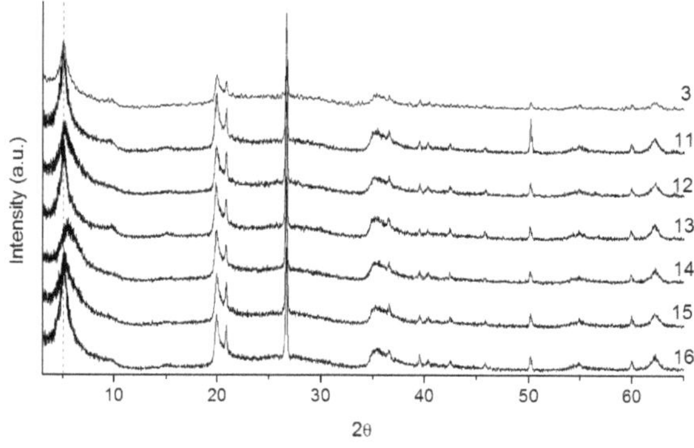

Figure 4. X-ray diffractograms of the series 3 samples.

Table 8. Modified parameters, basal spacings and surface areas of series 3 samples.

Method	Solution Concentrations	Relation Clay/Water (g/mL)	Mass of Clay	d_{001} (Å)	S_{BET} (m²/g)	Observation (Pillaring Agent)
Natural	-	-	-	15.1	58	-
Traditional	0.2 mol/L	1/100	3 g	17.8	234	6 days
3	0.6 mol/L	1/100	10 g	17.9	216	6 days
11	0.6 mol/L	1/100	10 g	17.6	237	60 °C, 24 h
12	1.2 mol/L	1/100	20 g	17.7	191	6 days
13	1.2 mol/L	1/100	20 g	17.5	251	60 °C, 24 h
14	1.5 mol/L	1/100	50 g	16.6	179	6 days
15	1.2 mol/L	1/100	50 g	17.6	197	6 days
16	1.2 mol/L	1/100	50 g	17.6	233	60 °C, 24 h

From the analysis of the data in Table 8, it could be noticed that when increasing the amount of pillared clay (methods 12, 14 and 15), a slight reduction of surface area occurred, with the lowest surface area obtained by method 14 (179 m²/g). For methods 11, 13, and 16, where the pillaring agent was synthesized in the course of a day with a heating stage at 60 °C, the surface areas remained high, even when pillaring 50 g of clay (Figure 5).

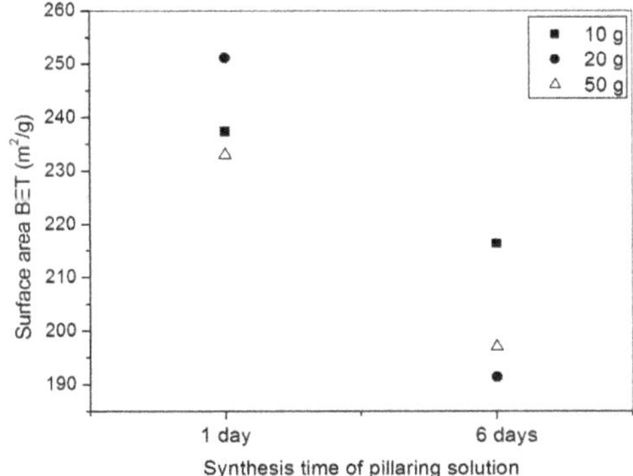

Figure 5. Values of surface area BET to several amounts of pillared clays as a function of time of preparation of the pillaring agent.

All pillared clays synthesized by the pillaring agent prepared in the course of 1 day (methods 11, 13 and 16) obtained surface areas larger than 200 m²/g. The clays pillared by the pillaring agent prepared over the course of 6 days obtained lower surface areas. Table 9 presents the data referring to the textural parameters from samples of this series. Not only was the surface area higher for samples pillared by the method using heating, other data such as total pore volume and micropore volume were larger for the samples pillared by this method.

Table 9. Textural parameters of series 3 samples.

Sample	S_{BET} (m²/g)	S_{micro} (m²/g)	S_{ext} (m²/g)	V_{total} (cm³/g)	V_{micro} (cm³/g)
Natural	58	19	39	0.070	0.010
Traditional	234	195	39	0.146	0.100
3	216	180	36	0.131	0.092
11	237	196	41	0.147	0.102
12	191	154	37	0.122	0.079
13	251	207	44	0.156	0.106
14	179	137	42	0.122	0.071
15	197	157	40	0.127	0.080
16	233	199	34	0.143	0.102

Comparing sample 15 (50 g of clay pillared by the traditional method) with sample 16 (50 g pillared by the method using heating), these two samples show the same basal spacing. However, a higher surface area was obtained for sample 16 (233 m^2/g); for sample 15, the surface area was 197 m^2/g. This lower value was most likely due to the pillaring agent utilized, where the concentration of the solution used was 1.2 mol/L, and even after 6 days of aging, the pillaring solution showed a little turbidity. Because method 16 used heating during the synthesis of the pillaring agent, no turbidity occurred, and the solution was completely clear when used, not presenting the problems reported in methods 14 and 15.

Therefore, the method of synthesis of the pillaring agent that applies heating at 60 °C with subsequent aging of the pillaring solution for 24 h at ambient temperature is the most suitable for pillaring large quantities of clay, which requires concentrated solutions. In method 16, fifty grams of pillared clay were produced by using this methodology. The basal spacing and surface area obtained (17.6 Å and 233 m^2/g) are in line with literature results achieved in pillaring procedures employing few grams of clay [14,36]. The traditional method is efficient just for small quantities of clay because increasing the concentration of the pillaring agent creates turbidity, generating other Al species.

4. Conclusions

Three series of experiments (water effect, method effect and increase in the amount of pillared clay) were studied in order to verify their influence in the final properties (basal spacing and surface area) of pillared clays. In series 1, diluted clay suspensions (minimum 1 g/50 mL) are required to obtain pillared materials with basal spacings and surface areas superior than 17 Å and 200 m^2/g, respectively.

In series 2, the most outstanding pillaring methodology employed was method 11, in which the pillaring agent was synthesized in 24 h with heating at 60 °C. The resulting Al-PILC presented basal spacing of 17.5 Å and surface area of 237 m^2/g.

In series 3, fifty grams of pillared clay with high basal spacing and surface area (17.6 Å and 233 m^2/g, respectively) were obtained by applying the best pillaring method studied in series 2 (pillaring agent synthesized at 60 °C with subsequent aging over 24 h at ambient temperature) and amount of water in clay suspension (1 g of clay/100 mL of water). The textural properties and basal spacing obtained by this material are comparable to the results obtained with the traditional pillaring method employing just 3 g of clay.

Acknowledgments: The authors gratefully acknowledge the Conselho Nacional de Desenvolvimento Científico e Tecnológico (CNPq) for financial support, the Federal University of Rio Grande do Norte (UFRN) for the physical structure and the Regional Integrated University (URI) for sample analysis.

Author Contributions: Francine Bertella and Sibele Pergher conceived and designed the experiments; Francine Bertella performed the experiments; Francine Bertella and Sibele Pergher analyzed the data and wrote the paper.

Conflicts of Interest: The authors declare no conflict of interest.

References

1. Kloprogge, T.T. Synthesis of Smectites and Porous Pillared Clay Catalysts: A Review. *J. Porous Mater.* **1998**, *5*, 5–41. [CrossRef]
2. Pergher, S.B.C.; Corma, A.; Fornes, V. Materiales laminares pilareados: Preparación y propiedades. *Quím. Nova* **1999**, *22*, 693–709. [CrossRef]
3. Mishra, T. Transition Metal Oxide-Pillared Clay Catalyst: Synthesis to Application. In *Pillared Clays and Related Catalysts*; Gil, A., Korili, S.A., Trujilano, R., Vicente, M.A., Eds.; Springer Science + Business Media: New York, NY, USA, 2010; pp. 99–128.
4. Occelli, M.L.; Bertrand, J.A.; Gould, S.A.C.; Dominguez, J.M. Physicochemical characterization of a Texas montmorillonite pillared with polyoxocations of aluminum Part I: The microporous structure. *Microporous Mesoporous Mater.* **2000**, *34*, 195–206. [CrossRef]

5. Kloprogge, T.T.; Evans, R.; Hickey, L.; Frost, R.L. Characterization and Al-pillaring of smectites from Miles, Queensland (Australia). *Appl. Clay Sci.* **2002**, *20*, 157–163. [CrossRef]

6. Yuan, P.; He, H.; Bergaya, F.; Wu, D.; Zhou, Q.; Zhu, J. Synthesis and characterization of delaminated iron-pillared clay with meso–microporous structure. *Microporous Mesoporous Mater.* **2006**, *88*, 8–15. [CrossRef]

7. Sivaiah, M.V.; Petit, S.; Brendlé, J.; Patrier, P. Rapid synthesis of aluminium polycations by microwave assisted hydrolysis of aluminium via decomposition of urea and preparation of Al-pillared montmorillonite. *Appl. Clay Sci.* **2010**, *48*, 138–145. [CrossRef]

8. Gil, A.; Assis, F.C.C.; Albeniz, S.; Korili, S.A. Removal of dyes from wastewaters by adsorption on pillared clays. *Chem. Eng. J.* **2011**, *168*, 1032–1040. [CrossRef]

9. Aznárez, A.; Korili, S.A.; Gil, A. The promoting effect of cerium on the characteristics and catalytic performance of palladium supported on alumina pillared clays for the combustion of propene. *Appl. Catal. A* **2014**, *474*, 95–99. [CrossRef]

10. Jalil, M.E.R.; Baschini, M.; Castellón, E.R.; Molina, A.I.; Sapag, K. Effect of the Al/clay ratio on the thiabendazol removal by aluminum pillared clays. *Appl. Clay Sci.* **2014**, *87*, 245–253. [CrossRef]

11. Pálinkó, I.; Lázár, K.; Kiricsi, I. Cationic mixed pillared layer clays: Infrared and Mössbauer characteristics of the pillaring agents and pillared structures in Fe, Al and Cr, Al pillared montmorillonites. *J. Mol. Struct.* **1997**, *410–411*, 547–550. [CrossRef]

12. Carriazo, J.; Guélou, E.; Barrault, J.; Tatibouët, J.M.; Molina, R.; Moreno, S. Catalytic wet peroxide oxidation of phenol by pillared clays containing Al-C-Fe. *Water Res.* **2005**, *39*, 3891–3899. [CrossRef] [PubMed]

13. Olaya, A.; Moreno, S.; Molina, R. Synthesis of pillared clays with Al_{13}-Fe and Al_{13}-Fe-Ce polymers in solid state assisted by microwave and ultrasound: Characterization and catalytic activity. *Appl. Catal. A* **2009**, *370*, 7–15. [CrossRef]

14. Sanabria, N.R.; Centeno, M.A.; Molina, R.; Moreno, S. Pillared clays with Al-Fe and Al-Ce-Fe in concentrated medium: Synthesis and catalytic activity. *Appl. Catal. A* **2009**, *356*, 243–249. [CrossRef]

15. Catrinescu, C.; Arsene, D.; Apopei, P.; Teodosiu, C. Degradation of 4-chlorophenol from wastewater through heterogeneous Fenton and photo-Fenton process, catalyzed by Al–Fe PILC. *Appl. Clay Sci.* **2012**, *58*, 96–101. [CrossRef]

16. Mnasri, S.; Frini-Srasra, N. Evolution of Brönsted and Lewis acidity of single and mixed pillared bentonite. *Infrared Phys. Technol.* **2013**, *58*, 15–20. [CrossRef]

17. Bertella, F.; Pergher, S.B.C. Pillaring of bentonite clay with Al and Co. *Microporous Mesoporous Mater.* **2015**, *201*, 116–123. [CrossRef]

18. Sterte, J. Hydrothermal treatment of hydroxycation precursor solutions. *Catalysis Today* **1988**, *2*, 219–231. [CrossRef]

19. Furrer, G.; Ludwig, C.; Schindler, P. On the Chemistry of the Keggin Al_{13} Polymer. Acid-Base Properties. *J. Colloid Interface Sci.* **1992**, *149*, 56–67. [CrossRef]

20. Schoonheydt, R.A.; Leeman, H.; Scorpion, A.; Lenotte, I.; Grobet, P. The Al pillaring of clays. Part II. Pillaring with $[Al_{13}O_4(OH)_{24}(H_2O)_{12}]^{7+}$. *Clays Clay Miner.* **1994**, *42*, 518–525. [CrossRef]

21. Pizarro, A.H.; Monsalvo, V.M.; Molina, C.B.; Mohedano, A.F.; Rodriguez, J.J. Catalytic hydrodechlorination of p-chloro-m-cresol and 2,4,6-trichlorophenol with Pd and Rh supported on Al-pillared clays. *Chem. Eng. J.* **2015**, *273*, 363–370. [CrossRef]

22. Aznárez, A.; Delaigle, R.; Eloy, P.; Gaigneaux, E.M.; Korili, S.A.; Gil, A. Catalysts based on pillared clays for the oxidation of chlorobenzene. *Catal. Today* **2015**, *246*, 15–27. [CrossRef]

23. Detoni, C.; Bertella, F.; Souza, M.M.V.M.; Pergher, S.B.C.; Aranda, D.A.G. Palladium supported on clays to catalytic deoxygenation of soybean fatty acids. *Appl. Clay Sci.* **2014**, *95*, 388–395. [CrossRef]

24. Sapag, K.; Mendioroz, S. Synthesis and characterization of micro-mesoporous solids: Pillared clays. *Colloids Surf. A* **2001**, *187–188*, 141–149. [CrossRef]

25. Schoonheydt, R.A.; Fynde, J.V.D.; Tubbax, H.; Leeman, H.; Stuyckens, M.; Lenotte, I.; Stone, W.E.E. The Al pillaring of clays. Part I. Pillaring with dilute and concentrated Al solutions. *Clays Clay Miner.* **1993**, *41*, 598–607. [CrossRef]

26. Kaloidas, V.; Koufopanos, C.A.; Gangas, N.H.; Papayannakos, N.G. Scale-up studies for the preparation of pillared layered clays at 1 kg per batch level. *Microporous Mater.* **1995**, *5*, 97–106. [CrossRef]

27. Fetter, G.; Heredia, G.; Velázquez, L.A.; Maubert, A.M.; Bosch, P. Synthesis of aluminum-pillared montmorillonites using highly concentrated clay suspensions. *Appl. Catal. A* **1997**, *162*, 41–45. [CrossRef]

28. Moreno, S.; Gutierrez, E.; Alvarez, A.; Papayannakos, N.G.; Poncelet, G. Al-pillared clays: From lab syntheses to pilot scale production Characterisation and catalytic properties. *Appl. Catal. A* **1997**, *165*, 103–114. [CrossRef]

29. Sánchez, A.; Montes, M. Influence of the preparation parameters (particle size and aluminium concentration) on the textural properties of Al-pillared clays for a scale-up process. *Microporous Mesoporous Mater.* **1998**, *21*, 117–125. [CrossRef]

30. Salerno, P.; Mendioroz, S. Preparation of Al-pillared montmorillonite from concentrated dispersions. *Appl. Clay Sci.* **2002**, *22*, 115–123. [CrossRef]

31. Olaya, A.; Moreno, S.; Molina, R. Synthesis of pillared clays with aluminum by means of concentrated suspensions and microwave radiation. *Catal. Commun.* **2009**, *10*, 697–701. [CrossRef]

32. Leite, S.Q.M.; Dieguez, L.C.; San Gil, R.A.S.; Menezes, S.M.C. Pilarização de esmectita brasileira para fins catalíticos. Emprego de argila pilarizada na alquilação de benzeno com 1-dodeceno. *Quím. Nova* **2000**, *23*, 149–154. [CrossRef]

33. Rouquerol, F.; Rouquerol, J.; Sing, K. *Adsorption by Powders and Porous Solids, Principles, Methodology and Applications*; Academic Press: Cambridge, MA, USA, 1999.

34. Bieseki, L.; Bertella, F.; Treichel, H.; Penha, F.G.; Pergher, S.B.C. Acid Treatments of Montmorillonite-rich Clay for Fe Removal Using a Factorial Design Method. *Mater. Res.* **2013**, *16*, 1122–1127. [CrossRef]

35. Aouad, A.; Mandalia, T.; Bergaya, F. A novel method of Al-pillared montmorillonite preparation for potential industrial up-scaling. *Appl. Clay Sci.* **2005**, *28*, 175–182. [CrossRef]

36. Zhu, J.; Wen, K.; Zhang, P.; Wang, Y.; Ma, L.; Xi, Y.; Zhu, R.; Liu, H.; He, H. Keggin-Al 30 pillared montmorillonite. *Microporous Mesoporous Mater.* **2017**, *242*, 256–263. [CrossRef]

 *materials*

Article

Reuse of Pillaring Agent in Sequential Bentonite Pillaring Processes

Francine Bertella and Sibele B. C. Pergher *

Laboratory of Molecular Sieves—LABPEMOL, Institute of Chemistry, Federal University of Rio Grande do Norte—UFRN. Av. Senador Salgado Filho, 3000, Lagoa Nova, University Campus, 59072-970 Natal-RN, Brazil; francinebertella@gmail.com
* Correspondence: sibelepergher@gmail.com; Tel.: +55-84-9413-5418

Academic Editor: Antonio Gil
Received: 27 April 2017; Accepted: 30 May 2017; Published: 27 June 2017

Abstract: This work describes the synthesis and characterization of pillared clays using a new pillaring method: the reuse of the pillaring solution. First, an Al pillared clay (PILC) was synthesized, and after filtration, the pillaring agent was stored and reused for an additional three pillaring procedures (P1, P2, and P3). The filtered pillaring solution was stored for one year and then reused for one additional pillaring procedure (P4). The samples were analyzed using XRD, N_2 physisorption measurements and chemical analysis (EDX). All of the samples exhibited basal spacings larger than 17 Å and BET surface areas greater than 160 m^2/g. After the P4 pillaring, the pillaring agent was precipitated with a Na_2SO_4 solution, and the resulting solid was analyzed using XRD and SEM. The results indicated that even after a total of five pillaring procedures, Al_{13} ions were still present in solution. Therefore, it is possible to reuse the pillaring solution four times and to even store the solution for one year, which is important from an industrial perspective.

Keywords: pillaring solution; Keggin ion; reutilization

1. Introduction

Pillaring is the process by which a layered compound is transformed into a thermally stable micro- and/or mesoporous material with retention of the layered structure. The obtained material is a pillared compound or a pillared layered solid. Pillared clays (PILCs) constitute a special class of pillared layered solids. The goal of the pillaring process is to introduce micro- and mesoporosity into clay minerals. This is achieved using a combination of a smectite, for example, and a pillaring agent via an ion-exchange reaction, in which a two-dimensional channel network is formed [1]. Heating the clays imparts stability to the pillared clay by promoting permanent bonding between the pillar and the layers. The resulting materials have small cavities and a large surface area. These properties, along with their low cost, make pillared clays ideal for use as alternative catalysts to zeolites [2].

There are numerous oligomeric cations that could be used as pillars, as well as Al, Zr, Ti, Cr, Fe, Ce and others [1,3,4]. These cations could be used individually [2,5–8] or mixed [9–17]. In addition, there is an enormous amount of work dealing with the use of Al-pillared clays as supports of metal nanoparticles (Pd and Pt, for example) for several catalytic applications [18–21]

In this sense, most of the studies published to date have focused on the use of the Al polyoxocation as a pillaring agent. Solutions containing this complex are prepared through forced hydrolysis, either by the addition of a base to $AlCl_3$ or $Al(NO_3)_3$ solutions, up to an OH/Al molar ratio of 2.5, or by dissolution of Al powder in $AlCl_3$ [22]. The most important cationic complex formed is most likely the polynuclear $[Al_{13}O_4(OH)_{24}(H_2O)_{12}]^{7+}$, which is the so-called Keggin or Al_{13} ion [23,24]. Although the method of preparation is clearly of prime importance, many other factors influence the formation of pure Al_{13} in solution, such as the nature and initial concentration of the reagents used, the degree of hydrolysis (OH/Al ratio), the rate of adding the reactants, the temperature, and the time and temperature of aging the hydrolyzed solutions [25].

Several methods and pillaring solutions, species of pillars and types of clays have been discussed in the literature. Nevertheless, most of these studies were performed using pillared clays prepared at the laboratory scale, usually in small amounts. However, some authors have performed experiments using concentrated clay dispersions [13,26–32]. All of these studies used Al_{13} as a pillaring agent or Al_{13} mixed with another cation, such as Fe. However, to our knowledge, none of the works reported in the literature reused the pillaring agent more than once. Thus, the purpose of this investigation is to reuse the pillaring agent in other pillaring procedures for scale-up studies.

2. Materials and Methods

2.1. Synthesis of the Aluminum Pillared Clay Minerals

The clay used in this study is a bentonite that contains montmorillonite as the dominant clay mineral. The first step of the pillaring process is to prepare the pillaring agent. For this purpose, 500 mL of a 0.2 mol/L sodium hydroxide solution (NaOH, Vetec Fine Chemicals, Duque de Caxias, Brazil) and 250 mL of a 0.2 mol/L aluminum chloride hexahydrate solution ($AlCl_3 \cdot 6H_2O$, Fluka Analytical, Sigma-Aldrich, St. Louis, MO, USA) were used. The sodium hydroxide solution was slowly added to the aluminum chloride solution dropwise, which was then maintained at room temperature under stirring for six days. A ratio of 0.05 moles of Al/g of dry clay and a molar ratio of OH/Al = 2 [33] were used.

The clay dispersion was prepared by adding 3 g of clay to 300 mL of distilled water, followed by stirring at room temperature for 2 h. After this expansion process step, the pillaring process itself was performed by adding the pillaring agent solution to the clay dispersion. The mixture was stirred for 2 h at room temperature such that the natural cations of the clay could be exchanged for the previously prepared oligomers, thereby affording an intercalated clay. The solid was separated by vacuum filtration, washed with distilled water until the test for chloride was negative, dried at 60 °C and subsequently calcined at 450 °C to generate the pillared clay.

During the filtration step, the pillaring agent was separated. The separated pillaring agent was stored in an amber bottle and subsequently reused in three other pillarings—called P1, P2 and P3—as shown in Table 1. After filtration of the P3 clay, the pillaring agent was again stored in an amber bottle, and it was stored for one year. After this period, the pillaring agent was reused for pillaring P4. All other pillaring steps were performed as described above.

Table 1. Data on pillaring with reuse of the pillaring agent.

Sample	Reuse of Pillaring Agent
PILC	—
P1	First time
P2	Second time
P3	Third time
P4	Fourth time (after one year)

2.2. Preparation of Al_{13} Sulfate

The methodology used was based on [13] and [23] and is described below. After pillaring P4, the pillaring agent filtrate was added to a Na_2SO_4 (Vetec Fine Chemicals, Duque de Caxias, Brazil) solution (500 mL of 0.1 mol/L) to precipitate the Keggin ions. The mixture was initially stirred for 2 h, followed by aging for 48 h. Then, the solid was vacuum filtered, washed with distilled water, and dried in an oven at 60 °C.

2.3. Material Characterization

The natural and pillared clays were characterized using XRD, textural analysis by N_2 adsorption, chemical analysis and scanning electron microscopy. X-ray diffraction analyses were performed on a Rigaku Desktop MiniFlex II apparatus (Rigaku, Tokyo, Japan) using CuKα radiation (λ = 1.54 Å). The analyses were performed over the range of 1.5° to 65° 2θ using an X-ray tube voltage of 30.0 kV and current of 15.0 mA, scan rate of 5°/min and step size of 0.05. The textural analyses of the samples by N_2 adsorption were conducted using the adsorption/desorption of nitrogen with a Nova 2200 E device (Quantachrome, Boynton Beach, FL, USA). The samples were pre-treated for 3 h at 300 °C under vacuum prior to the analyses. Then, the adsorption/desorption of N_2 at liquid nitrogen temperature was measured. Chemical analyses of the samples were performed using an EDX-720 energy dispersive X-ray fluorescence spectrometer (Shimadzu, Kyoto, Japan). Characterization by scanning electron microscopy was performed using an ESEM-XL30 instrument (Philips, Eindhoven, The Netherlands). The samples were coated with gold to avoid the appearance of charges on the surface that could lead to distortions in the images.

3. Results and Discussion

Figure 1 presents the X-ray diffraction patterns of the starting clay (natural) and of the pillared clays. The PILC sample refers to the first pillaring, and P1, P2, P3, and P4 refer to samples obtained from reuse of the pillaring agent. It was observed that all samples were pillared due to the increase in the interlayer spacing relative to that of the natural clay (Table 2). The starting clay had a basal spacing of 15.12 Å, which is an indication that Ca^{2+} ions are present as interlayer cations [34]. For the pillared samples, basal spacings of approximately 17.8 Å were obtained. This basal spacing is in agreement with those reported in the literature for pillaring with Al [5,13,35,36]. This value refers to the sum of the thickness of the lamella with the approximate size of polyoxocation $[Al_{13}]^{7+}$ (9 Å) [25], which confirms the inclusion of Al pillars between the clay lamellae.

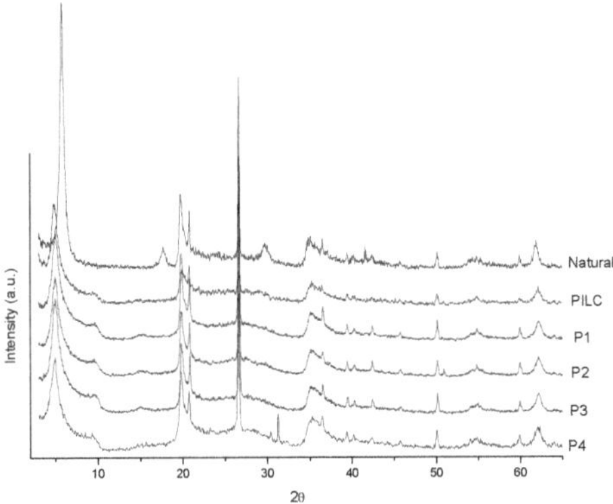

Figure 1. X-ray diffraction patterns of natural and pillared clays.

Table 2. Basal spacings obtained for the pillared clays and natural clay.

Samples	d_{001} (Å)
Natural	15.12
PILC	17.86
P1	17.86
P2	17.83
P3	17.54
P4	17.90

As shown in Figure 1, a shoulder also appeared on the peak at 9° 2θ, which corresponds to a basal spacing of approximately 9.6 Å. This spacing is characteristic of calcined clay, in which the layers are collapsed. This method of pillaring produces a pillared material, but some lamella may be collapsed.

The pillaring agent used in pillarings P1 to P4 was always the same; however, it was more dilute with each reuse. Nevertheless, the basal spacings of these samples were similar to that of the first pillared clay obtained using this pillaring agent (PILC). This result indicates that there are still Keggin ions (Al_{13}) present in the remaining solution, which may be due to two reasons: (1) at each pillaring, 100% of the cation-exchange capacity (CEC) of the clay is not being achieved and (2) the amount of Keggin ions present in the pillaring agent is considerably greater than the exchange capacity of the clay.

Using the following equation, we can calculate the mass of Keggin ions (theoretical) required to obtain 100% cation exchange using 3 g for the mass of clay used and 1.039 g/mmol for the molar mass (theoretical) of the Keggin ion $[AlO_4 Al_{12} (OH)_{24} \cdot (H_2O)_{12}]^{7+}$. The cation-exchange capacity of this clay is 1.79 mmol/g [37], but we have to take into account the stoichiometry of the reaction: Keggin ions have a charge +7 and Na ions have charge +1. Because of this, 0.255 mmol/g is the CEC correct value in this case.

$$\text{Mass} = \text{mass}_{clay} \times \text{CEC} \times \text{factor} \times \text{molar mass}$$

$$\text{Mass} = 3 \text{ g} \times 0.255 \text{ mmol/g} \times 1 \ (100\% \text{ CEC}) \times 1.039 \text{ g/mmol}$$

$$\text{Mass} = 0.79 \text{ g}$$

In theory, this mass is equivalent to 0.000765 moles of Keggin ions. For the synthesis of the pillaring agent, 0.05 moles of Al/g of clay are used. Because the majority of Al species present in solution are Keggin ions for the synthesis conditions for the pillaring agent used in this study [38,39], one can consider that, when using 0.05 moles, much of this amount is related to the number of moles of Keggin ions present in the pillaring agent. In fact, when using the above equation with a factor of 5 (in the five pillaring procedures that this pillaring agent was used, 100% cation exchange occurred), the number of moles of Keggin ions that would be required is 0.0038. Therefore, this amount of 0.05 moles of Al/g of clay was sufficient to perform five pillarings with 3 g of clay each.

Table 3 presents the data from the chemical analyses of the pillared clays and the starting clay. An increase in the Al content of the pillared samples relative to natural clay was observed, indicating that Al oligomers are being incorporated into the interlayer space of the clay. Notably, the Al content remained high even after one year (P4). Regarding the Si/Al ratio, a decrease in this ratio for pillared clays occurs in comparison to the natural clay, which confirms that the pillaring process occurred. It is also observed that the ratios obtained for samples P1 to P4 were similar, indicating that the pillaring agent can be reused four times. The Ca content in the pillared clays decreased relative to that of the natural clay, and in most samples, this element could not be detected. This was due to the cation exchange of Ca by oligomers. The Si content in the pillared clays was less than that in the starting clay. This difference is due to the aluminum inserted in the samples by changing the relative percentage. Regarding the values for Mg, the pillared clays exhibited a decrease compared to the natural clay, which indicates that part of the Mg ions also participate in the process and should be located in the interlayer space. The Fe content in the pillared clays increased relative to that in the natural clay, which may be associated with the technique because the EDX analysis is punctual.

Table 3. Chemical analyses of samples (%).

Sample	SiO_2	Al_2O_3	MgO	CaO_2	Fe_2O_3	Si/Al
Natural	69.96	19.08	5.61	2.73	2.62	3.52
PILC	59.37	33.87	3.59	n.d. [a]	3.17	1.69
P1	60.28	30.45	3.38	n.d.	5.89	1.90
P2	60.08	30.56	3.37	0.03	5.96	1.89
P3	60.15	30.53	3.18	n.d.	6.14	1.90
P4	59.86	30.76	3.36	n.d.	6.02	1.87

[a] n.d. = not detected.

To verify that the pillaring species was actually the Keggin ion, the pillaring solution of the P4 sample was filtered and then added to a solution of Na_2SO_4 to precipitate the remaining Keggin ions.

Figure 2 shows the XRD analyses of the solid precipitate (b) and of the Keggin ion synthesized using the same method before being used as a pillaring agent (a).

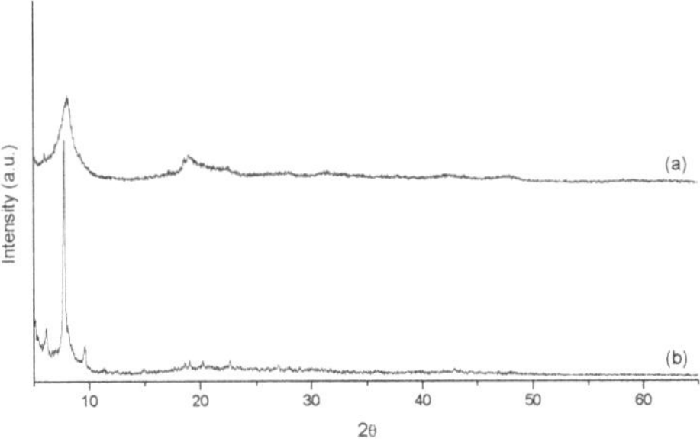

Figure 2. X-ray diffraction patterns of Keggin ion synthesized before being used as a pillaring agent (**a**) and after five pillaring procedures (**b**).

As shown in Figure 2, both samples exhibit the same diffraction peaks but with different intensities. Similar results were obtained by Tsuchida et al. [40], in which a solid precipitate from solutions of Al_{13} was obtained by adding sulfate salts with two types of crystals (I and II). The results of this study are similar to those of the type II crystals. Tsuchida et al. [40] also found that Keggin ion solutions synthesized using OH/Al ratios equal to 2 (the basicity used in this study) preferentially generate the type II crystals when precipitated with sulfate. Thus, it appears that even after five pillaring procedures, there are Keggin ions present in the pillaring agent. The method of precipitating Keggin ions with sulfate anions exerts a strong influence on the morphology of the solid Al_{13} [41]. Figure 3 presents scanning electron microscopy images of the precipitated Keggin ions before being used (a) and after five pillaring procedures (b).

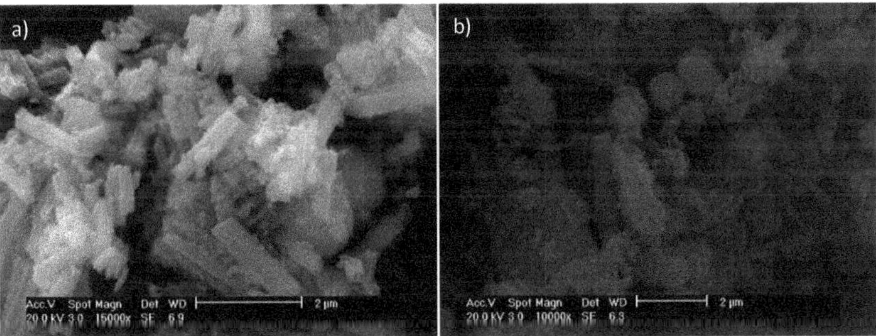

Figure 3. SEM micrographs of Keggin ion precipitates (**a**) before pillaring and (**b**) after five pillaring procedures.

Before pillaring (Figure 3a), the precipitated Keggin ions present a morphology similar to a fiberboard cylinder. After pillaring (Figure 3b), the precipitated Keggin ions exhibit different morphologies, with some particles possessing cylindrical shapes, others with spherical shapes and even amorphous agglomerates. Wang and Muhammed [41] found different morphologies (in the form of fibers, rectangles, and tetrahedrons) when precipitating the sulfate Keggin ion by changing only the precipitation time, and the tetrahedral morphology (as reported by [23]) was only obtained after long periods of precipitation, which demonstrates that the method employed to precipitate the Keggin ions exerts a great influence on the morphologies of the crystals.

The N_2 adsorption/desorption isotherms are shown in Figure 4. It was observed that all samples belong to the type IIb classification according to Rouquerol et al. [42], which is characteristic of materials that contain aggregates of plate-shaped particles, typical of clays. Regarding the type of hysteresis, all can be classified as type H3, which is characteristic of porous materials that consist of agglomerates of particles in the form of plates (lamellae) that produce slit-shaped pores [43].

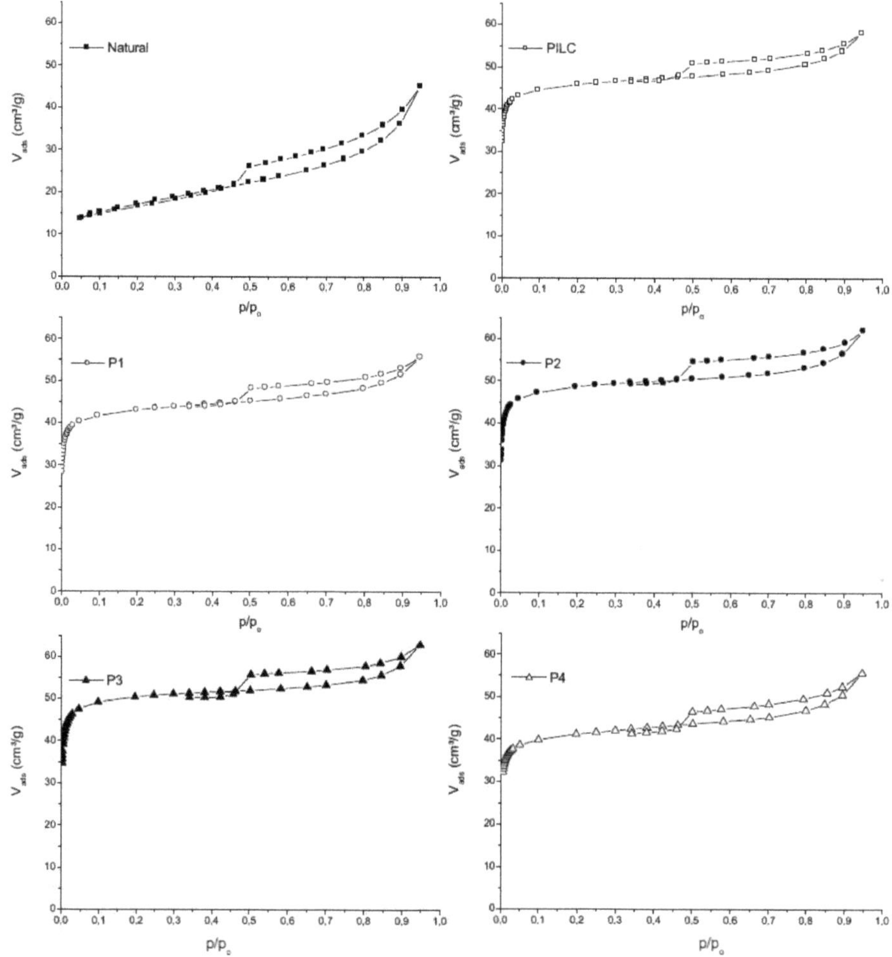

Figure 4. N_2 adsorption/desorption isotherms for natural clays, PILC, P1, P2, P3, and P4.

In comparing the isotherms of the pillared clays with the isotherm of the natural clay, it is observed that the treatment of natural clay with the Al oligomeric solution causes a considerable increase in the amount of adsorbed N_2. For $p/p_0 = 0.2$, the amount of N_2 adsorbed by the natural clay is 17.1 cm³/g, whereas for pillared clay P2 (the second time that the pillaring agent was reused), for example, the amount of N_2 adsorbed is 49 cm³/g, i.e., an approximately three-fold increase.

The isotherm data can be used to calculate the specific surface areas and pore volumes of the samples. The specific surface area was calculated using the BET method, and the total volume was calculated at a partial pressure of $p/p_0 = 0.95$. The area and volume of micropores were calculated using the t-plot method. The external surface is the difference between the BET surface area and micropore area, and it explains the contribution of meso- and macropores. These data are presented in Table 4.

Table 4. Textural parameters calculated from the adsorption isotherms for the prepared samples.

Sample	S_{BET} (m^2/g)	S_{micro} (m^2/g)	S_{ext} (m^2/g)	V_{micro} (cm^3/g)	V_{total} (cm^3/g)
Natural	58	19	39	0.010	0.070
PILC	183	171	12	0.066	0.090
P1	171	158	13	0,062	0.086
P2	194	183	11	0.071	0.096
P3	202	192	10	0.074	0.097
P4	164	150	14	0.058	0.086

A significant increase in the BET specific surface areas of the samples is observed after the pillaring, in which the BET areas of the pillared clays are approximately four times greater than the BET area of natural clay. Furthermore, note that the largest contribution to the high values of specific area are related to the increase in the area of micropores in the material, which is characteristic of pillared clays [33,44]. The total pore volume also increases following the pillaring procedure. With respect to the microporosity, there is an approximately 10-fold increase in microporosity in the pillared clays relative to that of the natural clay. Due to this increase in micropore volume, the micropore area of these materials is approximately 10 times greater than that of the natural clay.

The analysis of the results indicated that the specific surface areas of the samples that were pillared by reusing the pillaring agent from the PILC sample remained high, which demonstrates that the pillaring solution can be reused up to four times and can also be stored for long periods of time (P4).

4. Conclusions

The pillaring agent can be reused in other pillaring procedures without a loss in the primary characteristics of pillared clays, i.e., both the pillared clay obtained using the traditional method (PILC) and the pillared clays obtained by reusing the pillaring agent exhibited an increase in their basal spacings and high specific areas. Even the P4 sample, in which the pillaring agent was stored for one year before being reused for the fourth time, presented results that were as good as those for the clay pillared using the traditional method. Additionally, the XRD and SEM analyses revealed that there are Al_{13} ions present in solution, even after five pillarings, which makes the reuse of the pillaring agent possible.

Acknowledgments: The authors gratefully acknowledge the CNPq for financial support, the Federal University of Rio Grande do Norte (UFRN) for the physical structure and the Regional Integrated University (URI—Erechim) for the XRD and N_2 adsorption/desorption analyses.

Author Contributions: F.B. and S.B.C.P. conceived and designed the experiments; F.B. performed the experiments; F.B. and S.B.C.P. analyzed the data and wrote the paper.

Conflicts of Interest: The authors declare no conflict of interest.

References

1. Schoonheydt, R.A.; Jacobs, K.Y. Clays: From Two to Three Dimensions. In *Studies in Surface Science and Catalysis*; van Bekkum, H., Flanigen, E.M., Jacobs, P.A., Jansen, J.C., Eds.; Elsevier Science B.V.: Amsterdam, The Netherlands, 2001; pp. 299–343.
2. Kloprogge, J.T.; Evans, R.; Hickey, L.; Frost, R.L. Characterisation and Al-pillaring of smectites from Miles, Queensland (Australia). *Appl. Clay Sci.* **2002**, *20*, 157–163. [CrossRef]
3. Kloprogge, J.T.; Duong, L.V.; Frost, R.L. A review of the synthesis and characterization of pillared clays and related porous materials for cracking of vegetable oils to produce bilfuels. *Environ. Geol.* **2005**, *47*, 967–981. [CrossRef]
4. Pergher, S.B.C.; Corma, A.; Fornes, V. Materiales laminares pilareados: Preparación y propiedades. *Quim. Nova* **1999**, *22*, 693–709. [CrossRef]

5. Occelli, M.L.; Bertrand, J.A.; Gould, S.A.C.; Dominguez, J.M. Physicochemical characterization of a Texas montmorillonite pillared with polyoxocations of aluminum Part I: The microporous structure. *Microporous Mesoporous Mater.* **2000**, *34*, 195–206. [CrossRef]

6. Sivaiah, M.V.; Petit, S.; Brendlé, J.; Patrier, P. Rapid synthesis of aluminium polycations by microwave assisted hydrolysis of aluminium via decomposition of urea and preparation of Al-pillared montmorillonite. *Appl. Clay Sci.* **2010**, *48*, 138–145. [CrossRef]

7. Yuan, P.; He, H.; Bergaya, F.; Wu, D.; Zhou, Q.; Zhu, J. Synthesis and characterization of delaminated iron-pillared clay with meso–microporous structure. *Microporous Mesoporous Mater.* **2006**, *88*, 8–15. [CrossRef]

8. Zhu, J.; Wen, K.; Zhang, P.; Wang, Y.; Ma, L.; Xi, Y.; Zhu, R.; Liu, H.; He, H. Keggin-Al30 pillared montmorillonite. *Microporous Mesoporous Mater.* **2017**, *242*, 256–263. [CrossRef]

9. Carriazo, J.; Guélou, E.; Barrault, J.; Tatibouët, J.M.; Molina, R.; Moreno, S. Catalytic wet peroxide oxidation of phenol by pillared clays containing Al–Ce–Fe. *Water Res.* **2005**, *39*, 3891–3899. [CrossRef] [PubMed]

10. Catrinescu, C.; Arsene, D.; Teodosiu, C. Catalytic wet hydrogen peroxide oxidation of para-chlorophenol over Al/Fe pillared clays (AlFePILCs) prepared from different host clays. *Appl. Catal. B Environ.* **2011**, *101*, 451–460. [CrossRef]

11. Olaya, A.; Moreno, S.; Molina, R. Synthesis of pillared clays with Al$_{13}$-Fe and Al$_{13}$-Fe-Ce polymers in solid state assisted by microwave and ultrasound: Characterization and catalytic activity. *Appl. Catal. A Gen.* **2009**, *370*, 7–15. [CrossRef]

12. Pálinkó, I.; Lázár, K.; Kiricsi, I. Cationic mixed pillared layer clays: Infrared and Mössbauer characteristics of the pillaring agents and pillared structures in Fe, Al and Cr, Al pillared montmorillonites. *J. Mol. Struct.* **1997**, *410–411*, 547–550. [CrossRef]

13. Sanabria, N.R.; Centeno, M.A.; Molina, R.; Moreno, S. Pillared clays with Al–Fe and Al–Ce–Fe in concentrated medium: Synthesis and catalytic activity. *Appl. Catal. A Gen.* **2009**, *356*, 243–249. [CrossRef]

14. Timofeeva, M.N.; Khankhasaeva, S.T.; Chesalov, Y.A.; Tsybulya, S.V.; Panchenko, V.N.; Dashinamzhilova, E.T. Synthesis of Fe,Al-pillared clays starting from the Al,Fe-polymeric precursor: Effect of synthesis parameters on textural and catalytic properties. *Appl. Catal. B.* **2009**, *88*, 127–134. [CrossRef]

15. Timofeeva, M.N.; Panchenko, V.N.; Matrosova, M.M.; Andreev, A.S.; Tsybulya, S.V.; Gil, A.; Vicente, M.A. Factors affecting the catalytic performance of Zr,Al-pillared clays in the synthesis of propylene glycol methyl ether. *Ind. Eng. Chem. Res.* **2014**, *53*, 13565–13574. [CrossRef]

16. Bertella, F.; Pergher, S.B.C. Pillaring of bentonite clay with Al and Co. *Microporous Mesoporous Mater.* **2015**, *201*, 116–123. [CrossRef]

17. Ding, M.; Zuo, S.; Qi, C. Preparation and characterization of novel composite AlCr-pillared clays and preliminary investigation for benzene adsorption. *Appl. Clay Sci.* **2015**, *115*, 9–16. [CrossRef]

18. Detoni, C.; Bertella, F.; Souza, M.M.V.M.; Pergher, S.B.C.; Aranda, D.A.G. Palladium supported on clays to catalytic deoxygenation of soybean fatty acids. *Appl. Clay Sci.* **2014**, *95*, 388–395. [CrossRef]

19. Aznárez, A.; Delaigle, R.; Eloy, P.; Gaigneaux, E.M.; Korili, S.A.; Gil, A. Catalysts based on pillared clays for the oxidation of chlorobenzene. *Catal. Today* **2015**, *246*, 15–27. [CrossRef]

20. Pizarro, A.H.; Molina, C.B.; Rodriguez, J.J.; Epron, F. Catalytic reduction of nitrate and nitrite with mono- and bimetallic catalysts supported on pillared clays. *J. Environ. Chem. Eng.* **2015**, *3*, 2777–2785. [CrossRef]

21. Pizarro, A.H.; Molina, C.B.; Rodriguez, J.J. Decoloration of azo and triarylmethane dyes in the aqueous phase by catalytic hydrotreatment with Pd supported on pillared clays. *RSC Adv.* **2016**, *6*, 113820–113825. [CrossRef]

22. Kloprogge, J.T. Synthesis of Smectites and Porous Pillared Clay Catalysts: A Review. *J. Porous Mater.* **1998**, *5*, 5–41. [CrossRef]

23. Furrer, G.; Ludwig, C.; Schindler, P. On the Chemistry of the Keggin Al$_{13}$ Polymer. Acid-Base Properties. *J. Colloid Interface Sci.* **1992**, *149*, 56–67. [CrossRef]

24. Schoonheydt, R.A.; Leeman, H.; Scorpion, A.; Lenotte, I.; Grobet, P. The Al pillaring of clays. Part II. Pillaring with [Al$_{13}$O$_4$(OH)$_{24}$ (H$_2$O)$_{12}$]$^{7+}$. *Clays Clay Miner.* **1994**, *42*, 518–525. [CrossRef]

25. Bergaya, F.; Aouad, A.; Mandalia, T. Pillared Clays and Clay Minerals. In *Handbook of Clay Science*; Bergaya, F., Theng, B.K.G., Lagaly, G., Eds.; Elsevier Ltd.: Amsterdam, The Netherlands, 2006; pp. 393–421.

26. Fetter, G.; Heredia, G.; Velázquez, L.A.; Maubert, A.M.; Bosch, P. Synthesis of aluminum-pillared montmorillonites using highly concentrated clay suspensions. *Appl. Catal. A Gen.* **1997**, *162*, 41–45. [CrossRef]

27. Kaloidas, V.; Koufopanos, C.A.; Gangas, N.H.; Papayannakos, N.G. Scale-up studies for the preparation of pillared layered clays at 1 kg per batch level. *Microporous Mater.* **1995**, *5*, 97–106. [CrossRef]

28. Moreno, S.; Gutierrez, E.; Alvarez, A.; Papayannakos, N.G.; Poncelet, G. Al-pillared clays: From lab syntheses to pilot scale production. Characterisation and catalytic properties. *Appl. Catal. A. Gen.* **1997**, *165*, 103–114. [CrossRef]

29. Olaya, A.; Moreno, S.; Molina, R. Synthesis of pillared clays with aluminum by means of concentrated suspensions and microwave radiation. *Catal. Commun.* **2009**, *10*, 697–701. [CrossRef]

30. Salerno, P.; Mendioroz, S. Preparation of Al-pillared montmorillonite from concentrated dispersions. *Appl. Clay Sci.* **2002**, *22*, 115–123. [CrossRef]

31. Sánchez, A.; Montes, M. Influence of the preparation parameters (particle size and aluminium concentration) on the textural properties of Al-pillared clays for a scale-up process. *Microporous Mesoporous Mater.* **1998**, *21*, 117–125. [CrossRef]

32. Schoonheydt, R.A.; Van den Eynde, J.; Tubbax, H.; Leeman, H.; Stuyckens, M.; Lenotte, I.; Stone, W.E.E. The Al pillaring of clays. Part I. Pillaring with dilute and concentrated Al solutions. *Clays Clay Miner.* **1993**, *41*, 598–607. [CrossRef]

33. Pergher, S.B.C.; Sprung, R. Pilarização de uma argila brasileira com poliidroxications de alumínio: Preparação, caracterização e propriedades catalíticas. *Quim. Nova* **2005**, *28*, 777–782. [CrossRef]

34. Gomes, C.F. *Argilas: O Que São e Para Que Servem*; Fundação Galouste Gulbenkian: Lisboa, Portugal, 1988; p. 457.

35. Gil, A.; Assis, F.C.C.; Albeniz, S.; Korili, S.A. Removal of dyes from wastewaters by adsorption on pillared clays. *Chem. Eng. J.* **2011**, *168*, 1032–1040. [CrossRef]

36. Jalil, M.E.R.; Vieira, R.S.; Azevedo, D.; Baschini, M.; Sapag, K. Improvement in the adsorption of thiabendazole by using aluminum pillared clays. *Appl. Clay Sci.* **2013**, *71*, 55–63. [CrossRef]

37. Bertella, F.; Acorsi, M.; Bieseki, L.; Scherer, R.P.; Penha, F.G.; Pergher, S.B.C.; Lengler, H.C.M. *Proceedings of Determinação da Capacidade de Troca Catiônica em Argilas*; XVI Encontro de Química da Região Sul: Blumenau, SC, Brazil, 2008.

38. Baes, C.F.; Mesmer, R.E. *The Hydrolysis of Cations*; John Wiley and Sons: New York, NY, USA, 1976.

39. Sterte, J. Hydrothermal treatment of hydroxycation precursor solutions. *Catal. Today.* **1988**, *2*, 219–231. [CrossRef]

40. Tsuchida, T.; Kitamura, K.; Inagak, M. Formation of Crystalline Sulfates from Al_{13} Polymer Solutions: Effect of Washing on the Transformation of Type I to Type II Crystals. *J. Mater. Chem.* **1995**, *5*, 1233–1236. [CrossRef]

41. Wang, M.; Muhammed, M. Novel synthesis of Al_{13}-cluster based alumina materials. *Nanostruct. Mater.* **1999**, *11*, 1219–1229. [CrossRef]

42. Rouquerol, F.; Rouquerol, J.; Sing, K. *Adsorption by Powders and Porous Solids: Principles, Methodology and Applications*; Academic Press: London, UK, 1999.

43. Gregg, S.I.; Sing, K.S.W. *Adsorption, Surface Area and Porosity*; Academic Press: London, UK, 1982.

44. Storaro, L.; Lenarda, M.; Ganzerla, R.; Rinaldi, A. Preparation of hydroxyl Al and Al/Fe pillared bentonites from concentrated clay suspensions. *Microporous Mater.* **1996**, *6*, 55–63. [CrossRef]

MDPI
St. Alban-Anlage 66
4052 Basel
Switzerland
Tel. +41 61 683 77 34
Fax +41 61 302 89 18
www.mdpi.com

Materials Editorial Office
E-mail: materials@mdpi.com
www.mdpi.com/journal/materials

www.ingramcontent.com/pod-product-compliance
Lightning Source LLC
LaVergne TN
LVHW070844170726
843515LV00005B/570